Sanctions, Tourism, and Resilience

This timely and significant book examines how sanctions reshape the landscape of global tourism and how destinations subjected to these measures respond to isolation, economic strain, and stigma. Tourism is often imagined as a borderless industry, yet it remains deeply shaped by international politics. Sanctions expose this reality by disrupting flows of visitors, capital, and knowledge, forcing destinations to rethink their survival within a constrained global order.

The book investigates the strategies through which sanctioned destinations endure, adapt, and sometimes transform, highlighting the coping mechanisms, innovations, and long-term adjustments that enable them to sustain functionality and competitiveness within the global tourism market. By centering resilience, the book shifts the narrative from the impacts of sanctions to the responses and solutions, offering a novel perspective on how tourism systems operate under geopolitical constraints. This is supported with a wide range of comparative case studies of countries with long-standing experiences of sanctions.

This significant volume is of interest to policy makers, practitioners, destination management and marketing organizations, and students of crisis and politics who seek to understand how tourism endures amid geopolitical tension and how resilience itself becomes a space of creativity, negotiation, and survival.

Siamak Seyfi is an associate professor in the Geography Research Unit at the University of Oulu, Finland. His research investigates the political and spatial dimensions of tourism, paying particular attention to questions of resilience, sustainability, and the ways tourism is entangled with geopolitics, especially within the MENA region.

Rasoul Afsari is an assistant professor of urban planning at Supreme National Defense University, Iran, and his work examines urban resilience, tourism planning, and sustainable urban development. He has published widely on how cities respond to crises and long-term pressures, with an interest in planning approaches that support more adaptive and equitable urban futures.

Ali Hosseini is an associate professor of urban geography in the Department of Human Geography and Planning at the University of Tehran, Iran. His research interests span policy analysis, spatial inequality, and tourism planning, with expertise in both quantitative and qualitative methods.

Contemporary Geographies of Leisure, Tourism and Mobility

Series Editor: C. Michael Hall

Dean's Chair in Marketing, Sustainability and Society, Massey Business School, Te Kura Whai Pakihi, Massey University, Albany, New Zealand

The aim of this series is to explore and communicate the intersections and relationships between leisure, tourism and human mobility within the social sciences.

It will incorporate both traditional and new perspectives on leisure and tourism from contemporary geography, e.g. notions of identity, representation and culture, while also providing for perspectives from cognate areas such as anthropology, cultural studies, gastronomy and food studies, marketing, policy studies and political economy, regional and urban planning, and sociology, within the development of an integrated field of leisure and tourism studies.

Also, increasingly, tourism and leisure are regarded as steps in a continuum of human mobility. Inclusion of mobility in the series offers the prospect to examine the relationship between tourism and migration, the sojourner, educational travel, and second home and retirement travel phenomena.

Routledge Studies in Contemporary Geographies of Leisure, Tourism and Mobility is a forum for innovative new research intended for research students and academics, and the titles will initially be available in hardback only. Titles include:

The Ethics of Volunteering in Tourism
A Critical Evaluation
Konstantinos Tomazos

Tourism Technology, Sustainability, and Local Community Empowerment
Edited by Hafizullah Dar, Kuldeep Singh, and Babu George

Sanctions, Tourism, and Resilience
Adaptation and Survival
Siamak Seyfi, Rasoul Afsari, and Ali Hosseini

For more information about this series, please visit: www.routledge.com/Contemporary-Geographies-of-Leisure-Tourism-and-Mobility/book-series/SE0522

Sanctions, Tourism, and Resilience

Adaptation and Survival

Siamak Seyfi, Rasoul Afsari, and Ali Hosseini

Routledge
Taylor & Francis Group
LONDON AND NEW YORK

First published 2026
by Routledge
4 Park Square, Milton Park, Abingdon, Oxon OX14 4RN

and by Routledge
605 Third Avenue, New York, NY 10158

Routledge is an imprint of the Taylor & Francis Group, an informa business

British Library Cataloguing-in-Publication Data
A catalogue record for this book is available from the British Library

ISBN: 978-1-041-10287-8 (hbk)
ISBN: 978-1-041-10289-2 (pbk)
ISBN: 978-1-003-65433-9 (ebk)

DOI: 10.4324/9781003654339

Typeset in Times New Roman
by Apex CoVantage, LLC

Contents

Figures

Tables

Boxes

Preface

Tourism has always been bound to politics, even if this connection is often disguised behind images of leisure and hospitality. Who travels, on what terms, and to which destinations is never a neutral process but one shaped by regulations, institutions, and international power relations. From passport regimes and airline agreements to financial infrastructures and security protocols, the movement of people is always conditioned by political choices. These structures may fade into the background in times of stability, but moments of rupture bring them to the forefront. This book emerges from the recognition that tourism is not only an economic sector but also a field where questions of power, justice, and resilience converge. By focusing on sanctions as a tool of statecraft, the chapters examine how coercive measures reshape mobility, stigmatize places, and generate new vulnerabilities. They also highlight the creativity and endurance of communities, businesses, and states that must cope with conditions of scarcity and prolonged uncertainty. Resilience runs as a central thread throughout the text, yet it is approached critically rather than romantically. The aim is not to celebrate recovery at all costs but to ask who becomes resilient, who pays the price, and what kinds of futures are sustained under sanctions. In doing so, the book seeks to widen the lens of both tourism studies and international relations, bringing them into closer conversation around the politics of mobility, survival, and adaptation. The chapters are intended to stimulate reflection and debate. They provide a foundation for further inquiry into the entanglements between tourism and geopolitics, but they also invite readers to think about broader questions: What does it mean to sustain an industry under duress? How do societies balance openness and security? And how might tourism be reimagined in a world where crises have become the norm rather than the exception? As such, we hope that, if only in some small way, this book contributes to a broader and more critical understanding of the geopolitics of tourism and of the ways sanctions as a unique type of crisis reveal the fragile intersections of mobility, power, and survival.

Siamak: This book is also the outcome of my own journey through the study of tourism under conditions of crisis. Writing it has meant confronting uncomfortable truths about how deeply politics penetrates what is often imagined as a sphere of leisure. It has also meant listening to the stories of individuals, businesses, and communities whose daily lives are marked by scarcity, uncertainty, and resilience. I dedicate this work to them and to all those whose livelihoods in tourism are caught in the crossfire of geopolitical disputes. Their persistence, creativity, and refusal to disappear from the global map remain the strongest reminder of why the study of tourism matters.

We wish to gratefully acknowledge the help and support of Emma Travis, Sarah Lewis, Harriet Cunningham, Lydia Kessell, and all at Routledge for their continuing support.

1 Sanctions in global politics and international relations

Introduction and background

Sanctions have become one of the defining instruments of twenty-first-century geopolitics, yet their consequences for tourism remain poorly understood until recently (e.g., Pratt & Alizadeh, 2018; Seyfi & Hall, 2019, 2020a, 2020b, 2020c; Seyfi et al., 2025). Headlines regularly announce new restrictions, and these measures ripple far beyond diplomacy and trade. This trend intensified under President Trump's leadership, as his administrations turned repeatedly to sanctions and tariffs as instruments of foreign policy. During his second presidency, Donald Trump triggered a global trade war by enacting steep tariffs on nearly all goods imported into the United States, driving tariff levels to historic highs and reshaping global economic relations (York & Durante, 2025). Tourism was not immune: data showed sharp declines in cross-border mobility between the United States and Canada as rising tariffs and hostile rhetoric altered perceptions of the United States as a welcoming destination (Wile, 2025). Neighboring states such as Canada recorded fewer cross-border trips with the United States as residents redirected their travel to alternative markets, while the European Union's sanctions after Russia's invasion of Ukraine disrupted established flows through closed airspace, blocked carriers, and restrictions on Russian tourists (Pala, 2025). These developments show that sanctions, once considered peripheral to tourism, now function as structural forces that reshape mobility patterns, alter destination competitiveness, and transform perceptions of accessibility in ways that reach deep into the organization of global travel and destination economies (Seyfi & Hall, 2019, 2020a).

Rather than isolated disruptions, sanctions now operate as enduring geopolitical conditions that steadily reshape tourism systems. They can affect who is able to travel, how financial transactions are routed, whether airlines and cruise operators can function, and how destinations are positioned in global imaginaries. For many destinations, sanctions constitute crises that are externally imposed, prolonged, and politically charged, unlike bounded shocks such as pandemics, natural disasters, or terror attacks. The challenge

DOI: 10.4324/9781003654339-1

is not rapid recovery but the capacity to continue operating under sustained constraint, which places resilience at the center of concern. Tourism workers, firms, and communities mobilize a range of strategies, some formalized, others improvised, to survive and adapt (Khodadadi, 2025; Seyfi et al., 2025). Yet such resilience is uneven and contested. What states present as national self-sufficiency may obscure the hardship borne by households and workers, particularly women and those in insecure or informal roles (Seyfi et al., 2025). Resilience must, therefore, be understood not as a neutral or universally positive process but as one shaped by geopolitical hierarchies, institutional weaknesses, and unequal distributions of risk and reward.

It is against this backdrop that this book positions sanctions as a distinct and understudied form of crisis in tourism studies. Whereas much of the existing literature treats crises as temporary disruptions with a clear trajectory of recovery (e.g., Mair et al., 2016; Beirman, 2020; Zhang & Zhu, 2024), sanctions impose enduring uncertainty and systemic strain. Understanding them requires attention both to their material effects, such as declining international arrivals, disrupted aviation networks, and exclusion from financial markets, and to the political dimensions through which resilience is defined and practiced.

Building on this foundation, the arguments developed here move beyond a narrow view of resilience as technical bouncing back. In sanctioned destinations, resilience takes multiple forms: small businesses cultivating domestic markets, informal networks sustaining basic operations, and state-led narratives of defiance. These practices are rarely straightforward. They are uneven, contested, and often contradictory, as strategies that preserve tourism may simultaneously entrench dependency, reinforce inequality, or generate new vulnerabilities.

The aim, therefore, is not only to examine how destinations survive under sanctions but also to ask what kinds of tourism futures are enabled or constrained by these prolonged and externally imposed disruptions. By examining destinations that have experienced sanctions, this book situates these measures within broader debates on tourism crises and resilience, advancing theoretical perspectives while grounding the discussion in lived realities. At the same time, it offers practical insights for destinations and businesses grappling with the effects of sanctions, highlighting both the possibilities for adaptation and the risks of intensified vulnerability.

The chapter proceeds as follows. It begins by clarifying definitions and examining what counts as a sanction, since the term has been applied to a wide range of measures that differ in scope, intent, and impact. It then examines the historical development of sanctions, showing how their use has expanded from narrow trade embargoes to more complex instruments of financial, diplomatic, and mobility restriction. The discussion then turns to the scales and forms of sanctions, highlighting the different levels at which they operate – international, national, and sectoral – and the ways in which they reshape tourism through restrictions on travel, investment, and reputation.

Definitions: what counts as a sanction?

There is no single authoritative definition of sanctions across international regimes. In general, they are understood as policy instruments designed to restrict customary interactions with a named target in order to pressure a change in behavior (van Bergeijk, 2021; Jentleson, 2022). Such measures may cut across trade, finance, investment, transport, and mobility, but they increasingly extend into cultural, educational, and sporting domains as well (Hufbauer et al., 2008; Cortright & Lopez, 2018). At their core, sanctions are coercive acts that seek to impose costs without resorting to direct military force. International sanctions are usually described in terms of foreign policy actions taken in response to a perceived wrongdoing by a state, organization, individual, or other actor. As Rossignol (1996) observed, these measures often include bans on the sale and shipment of products and restrictions on the purchase of exports, but may also involve non-commercial domains such as cultural exchanges, sports, and person-to-person interactions like the denial of visas or other travel documents (Box 1.2). Building on this, Crawford (1999, p. 5), in a widely cited study of apartheid South Africa, defined sanctions as "the denial of customary interactions (strategic, economic, or social); they are intended to promote social, political, or economic change in a target state. As the denial of customary interactions, sanctions speak louder than words: imposition of sanctions communicates the threat of more sanctions and also the promise of release from embargoes if the target meets certain conditions."

The terminology surrounding sanctions further reflects this diversity. *Embargo* usually refers to restrictions on the export of a particular good or category of goods. For instance, the U.S. arms embargo on China, imposed after the Tiananmen Square crackdown in 1989, continues to prevent the transfer of military-related equipment (Hu, 2016). Broader *sanctions* may combine trade prohibitions, import restrictions, financial measures, and mobility controls, as in the comprehensive UN regime against Iraq in the 1990s, which restricted oil exports and limited access to global financial markets (Malone, 2006). *Tariffs* are also frequently identified as a sanctioning tool, despite their presentation as trade measures. As Jenkins (2025) argues, tariffs represent "sanctions by another name," functioning as economic penalties that can be as disruptive as formal prohibitions, with the President Trump administration's tariffs on Syria serving as a striking example (Box 1.1). By contrast, *boycotts* are often initiated outside formal state channels, involving consumer action, institutional divestment, or collective refusal to engage with a target. The anti-apartheid boycotts of South Africa illustrate this, as cultural, academic, and sporting organizations played a role in isolating the regime even before comprehensive state sanctions were enacted (Nixon, 2016).

Box 1.1 Sanctions, tariffs, and the limits of economic coercion

In recent debates, tariffs have been increasingly framed as a form of sanction, raising questions about their effectiveness and legitimacy as tools of foreign policy. Writing in *The Guardian*, Simon Jenkins (2025) argues that tariffs are "sanctions by another name," deployed as punitive economic measures that often harm populations more than regimes. He highlights the Trump administration's imposition of a 41% tariff on Syrian exports, a move framed as retaliation against alleged economic misconduct. Yet Jenkins contends that the real barrier to Syria's recovery has been the longstanding Western sanctions regime, which has restricted trade, finance, and investment for more than a decade. These measures, he suggests, contributed to mass displacement, entrenched poverty, and the consolidation of Assad's power, while failing to deliver meaningful political change. Extending this critique, Jenkins situates sanctions within a broader pattern of what he calls "neo-imperial outreach," where economic coercion substitutes for direct military intervention. He notes that the number of U.S. sanctions has risen from around 900 in 2000 to more than 15,000 today, with the UK also maintaining thousands of restrictions through an extensive bureaucratic apparatus. Far from destabilizing authoritarian regimes, such sanctions have often coincided with their persistence, with North Korea, Iran, Russia, and Cuba among the most heavily sanctioned yet enduring states. The case of Russia illustrates this paradox. Despite some of the most sweeping sanctions in modern history, Russia's economy has continued to grow, unemployment has remained low, and its trade has been redirected through partners such as China and Kazakhstan. Jenkins concludes that sanctions frequently serve as symbolic gestures of political resolve rather than effective instruments of change, entrenching hardship without altering the policies they are meant to target.

Source: Jenkins (2025)

Historical development of sanctions

Sanctions are often presented as a distinctly modern policy tool, yet their lineage stretches back centuries, embedded in the long history of statecraft and coercive diplomacy (van Bergeijk, 2021; Jentleson, 2022). Far from being novel, restrictions on trade, mobility, and interaction have repeatedly been employed as a means of exerting political pressure without recourse to outright war. What has changed over time is not the basic principle of denial but the institutional forms, justifications, and instruments through which sanctions are enacted (Rodríguez, 2024; Mulder, 2022; van Bergeijk, 2021). Early examples reveal their deep historical roots. In classical Greece, the Megarian decree barred merchants of Megara from Athenian markets, using economic exclusion as a weapon of rivalry. By the early twentieth century, such measures were increasingly framed within the legal and normative language of international order (Hufbauer, 1998). The League of Nations promoted economic sanctions as an alternative to armed conflict, envisioning collective economic pressure as a deterrent against aggression. Yet enforcement remained uneven. The League's sanctions against Italy during its invasion of Ethiopia in the 1930s, for instance, excluded oil, the most critical commodity, rendering the measures ineffective and exposing the limits of collective resolve (Hufbauer, 1998, Hufbauer et al., 2008).

The Second World War and its lead-up further blurred the boundary between sanctions and acts of war. U.S. embargoes on Japan in 1940–1941, covering aviation fuel and other strategic goods, were read in Tokyo not as limited coercion but as existential threats, contributing to the escalation that culminated in Pearl Harbor. Such episodes highlight how sanctions, while formally non-military, can generate severe strategic consequences (Hufbauer et al., 2008; Mulder, 2022; Jentleson, 2022). During the Cold War, sanctions became more selective and symbolic, constrained by Security Council gridlock yet significant in certain contexts. Sanctions against Rhodesia and apartheid South Africa stand out, combining trade restrictions with sporting and cultural boycotts that carried strong normative weight. These cases marked an important expansion of the sanctioning repertoire, showing how exclusion from international cultural and symbolic arenas could reinforce political isolation (Jentleson, 2022; Hufbauer et al., 2008). The 1990s marked a turning point, as sanctions shifted from being occasional instruments to a policy default. The comprehensive regime against Iraq following its invasion of Kuwait demonstrated the scale and ambition of post-Cold War sanctions, but it also triggered fierce debates over humanitarian impacts. The sweeping trade and financial restrictions caused significant civilian suffering, raising questions about the legitimacy of comprehensive sanctions as tools of governance (Cortright & Lopez, 1999). From the 2000s to the present, sanctions have become increasingly targeted and financialized. The rise of asset freezes, travel bans, export controls, and restrictions linked to global payment systems reflects both technological change and the growing centrality of finance in global governance (Box 1.2). The development of compliance industries, centered on banks, insurers, and digital payment providers, has elevated the extraterritorial reach of sanctions, making them harder for targeted states and businesses to circumvent. Today, sanctions extend across domains, from banking to aviation, from cultural exchange to digital communication, illustrating their transformation into a routine and systemic feature of international politics. Table 1.1 summarizes this historical trajectory, tracing the evolution of rationales, instruments, and implementation practices across successive eras.

Box 1.2 Travel bans as a form of sanction

Travel bans are a common non-economic sanction designed to restrict mobility by preventing listed individuals from entering or transiting through certain territories. They are often combined with asset freezes, signaling international condemnation while limiting access to resources and networks. In September 2025, the European Commission considered extending these measures to ordinary civilians by proposing a full ban on Russian tourists as part of its 19th sanctions package against Moscow. The initiative, led by Poland, Finland, and the Baltic states, would mark an unprecedented expansion of travel restrictions within the EU. Supporters argued that the surge in Russian visitors, over half a million Schengen visas were issued in 2024, undermined the bloc's credibility while the Kremlin continued its war in Ukraine. "We can't just accept Russians travelling and enjoying their lives while their government is killing Ukrainians," one EU diplomat stated (Euractiv, 2025).

Yet the proposal exposed divisions among member states. Tourism-dependent economies such as Italy, Spain, and Greece expressed reluctance, while Hungary and Slovakia signaled resistance, raising doubts about the unanimous approval required for adoption. The debate highlighted a broader tension: whether travel bans should remain targeted at elites and officials or extend to ordinary citizens, with direct consequences for tourism flows, host economies, and the politics of mobility in wartime Europe.

Source: Brzozowski (2025)

Table 1.1 The evolution of sanctions

Era	*Main rationale claimed*	*Typical tools*	*Implementation features*
Pre-1914	Wartime auxiliary	Blockades, contraband rules	Customary maritime law
1919–1939	Peace through economic pressure	Comprehensive trade/finance bans	League authorizations, weak enforcement
1945–1989	Selective multilateralism	Arms embargoes, limited trade bans	Security Council gridlock; normative campaigns
1990s	Sanctions as policy default	Comprehensive regimes, early financial tools	Iraq model; humanitarian costs debates
2000s–present	Targeted design; financial centrality	Listings, asset freezes, SWIFT-related measures, export controls, travel bans	Extraterritorial reach; compliance industry

Source: After Hufbauer, 1998; Farrall, 2007; Hufbauer et al., 2008; Mulder, 2022; Jentleson, 2022

Scales and forms of sanctions

Sanctions are not a uniform instrument but operate across multiple scales of governance and through a wide range of measures. They may be imposed multilaterally by international organizations, regionally by supranational bodies, unilaterally by individual states, or initiated by non-state actors. Each scale carries its own legal basis, authority, and reach, producing uneven consequences for both sanctioned targets and the wider global system (van Bergeijk, 2021). At the *multilateral level*, the United Nations Security Council authorizes sanctions under Chapter VII of the UN Charter in response to threats to international peace and security. These measures are legally binding on all member states, though their enforcement depends on national capacity and political will (United Nations, 2025). UN sanctions commonly include arms embargoes, asset freezes, and travel bans. For tourism, this may mean restrictions on airlines deemed unsafe or linked to sanctioned entities, as well as visa denials for listed individuals.

Regional organizations have also become powerful sanctioning actors. The European Union, through its Common Foreign and Security Policy (CFSP), operates one of the most extensive sanctions systems globally (European Commission, 2025). Its measures include sectoral bans, targeted listings of individuals and entities, and export controls. The EU's central role as a travel market and financial regulator elevates these sanctions: they affect payment systems, tour operators, and airlines, while package travel firms must vet partners for exposure to sanctioned markets. Recent sanctions on Russia illustrate this reach, from aviation restrictions to constraints on financial transactions linked to tourism (Pala, 2025).

At the *unilateral level*, states utilize sanctions through domestic statutes or executive orders. The United States is the most influential actor in this respect. Its dominance in global finance, underpinned by the centrality of the U.S. dollar and the authority of the Office of Foreign Assets Control (OFAC), ensures that unilateral U.S. measures often have global consequences. The risk of secondary exposure pressures non-U.S. firms, banks, and insurers to withdraw from sanctioned markets to avoid penalties (Moiseienko, 2021). For tourism, this has meant the suspension of international card networks in Iran, the blocking of global distribution systems (GDS) in Cuba, and the cancellation of aircraft leasing contracts in Russia after the 2022 invasion of Ukraine.

In addition to these state-led measures, *non-state actors* also employ sanctioning practices, albeit within the limits set by national legal frameworks. The Boycott, Divestment and Sanctions (BDS) movement against Israel is one prominent example (Box 1.3). Operating outside formal state mandates, it mobilizes consumer boycotts, institutional divestment, and cultural campaigns to pressure political change (BDS Movement, 2025). Such initiatives frequently generate countermeasures, including U.S. state-level legislation restricting public institutions from engaging with BDS activities.

The breadth of sanctioning practices also extends beyond trade and finance. Economic restrictions often dominate headlines, yet non-economic measures, ranging from travel bans to cultural boycotts, play a significant role in shaping international interactions. These measures affect tourism in immediate ways, limiting mobility, altering perceptions of accessibility, and redefining the boundaries of cultural exchange. Table 1.2 outlines the main categories of sanctions, highlighting both economic and non-economic tools and illustrating how sanctions now span multiple domains of global life. Moreover, Table 1.3 presents a comparative overview of countries currently subject to international sanctions. It lists the positions of key sanctioning authorities: the Office of Foreign Assets Control (US), the United Nations, the European Union, the United Kingdom, Canada, Australia, Japan, and China. While most of these states adopt a broadly aligned stance, variations are visible. For instance, the United States maintains sanctions against Cuba, whereas many others do not.

Table 1.2 Scales and forms of sanctions in contemporary international politics

Category	*Sub-type / instrument*	*Examples of measures*
Economic and financial sanctions	Trade embargoes	Prohibition of exports or imports of specific goods (e.g., arms embargoes, oil bans).
	Comprehensive trade bans	Restrictions covering most or all goods and services exchanged with a target state.
	Financial restrictions	Freezing of assets, suspension of credit, limits on access to international banking systems.
	Currency and capital controls	Restrictions on currency transactions, prohibition on investment flows.
	Sectoral sanctions	Targeting key industries such as energy, defense, or technology.
	Extraterritorial/ secondary sanctions	Measures penalizing third parties for dealings with the sanctioned target.
Non-economic sanctions	Travel bans	Denial of visas or entry to individuals associated with targeted regimes or activities.
	Diplomatic sanctions	Downgrading or severing of diplomatic relations; withdrawal of recognition.
	Cultural and sporting restrictions	Bans on participation in international cultural, academic, or sporting events.
	Arms embargoes (when framed separately)	Prohibition on the sale or transfer of weapons and military-related technology.
	Information and communication restrictions	Suspension of broadcasting rights, media bans, internet access limitations.

Source: After Hufbauer, 1998; Hufbauer et al., 2008; Farrall, 2007; Cortright & Lopez, 1999, 2018; Mulder, 2022; Jentleson, 2022

Table 1.3 International sanctions by country and sanctioning authority

Country	*OFAC (US)*	*UN*	*EU*	*UK*	*Canada*	*Australia*	*Japan*	*China*
Afghanistan	Yes	Yes	Yes	Yes	Yes	Yes	Yes	No
Belarus	Yes	Yes	Yes	Yes	Yes	Yes	Yes	No
Burundi	No	Yes	Yes	Yes	Yes	Yes	No	No
Central African Republic	No	Yes	Yes	Yes	Yes	Yes	No	No
China	Yes	No	No	No	Yes	?	Yes	-
Cuba	Yes	No	No	No	No	No	No	No
Democratic Republic of Congo	No	Yes	Yes	Yes	Yes	Yes	No	No
Eritrea	No	Yes	Yes	Yes	Yes	Yes	No	No
Guinea-Bissau	No	Yes	Yes	Yes	Yes	Yes	No	No
Iran	Yes	Yes	Yes	Yes	Yes	Yes	Yes	No
Iraq	Yes	Yes	Yes	Yes	Yes	Yes	No	No
Lebanon	Yes	No	Yes	Yes	Yes	Yes	No	No
Libya	Yes	Yes	Yes	Yes	Yes	Yes	No	No
Mali	No	Yes	Yes	Yes	Yes	Yes	No	No
Myanmar (Burma)	Yes	Yes	Yes	Yes	Yes	Yes	Yes	No
Nicaragua	Yes	No	Yes	Yes	Yes	Yes	No	No
North Korea	Yes	Yes	Yes	Yes	Yes	Yes	Yes	Yes
Russia	Yes	No	Yes	Yes	Yes	Yes	Yes	Yes
Somalia	Yes	Yes	Yes	Yes	Yes	Yes	No	No
South Sudan	Yes	Yes	Yes	Yes	Yes	Yes	No	No
Sudan	Yes	Yes	Yes	Yes	Yes	Yes	No	No
Syria	Yes	Yes	Yes	Yes	Yes	Yes	No	No
Ukraine (Crimea/Donetsk/ Luhansk)	Yes	No	Yes	Yes	Yes	Yes	Yes	No
Venezuela	Yes	No	Yes	Yes	Yes	Yes	No	No
Yemen	Yes	Yes	Yes	Yes	Yes	Yes	No	No
Zimbabwe	Yes	Yes	Yes	Yes	Yes	Yes	No	No

Source: United Nations, 2025; European Commission, 2025; EU Sanctions Map (https://www.sanctionsmap.eu/#/main), OFAC, 2025

Box 1.3 The Boycott, Divestment and Sanctions (BDS) movement

The Boycott, Divestment and Sanctions (BDS) movement was launched in 2005 by a coalition of over 170 Palestinian civil society organizations, including unions, refugee groups, women's associations, and professional networks. It calls for non-violent pressure on Israel until it complies with international law by meeting three central demands: an end to occupation and settlement expansion, equal rights for Palestinian citizens of Israel, and recognition of the right of return for Palestinian refugees as outlined in UN Resolution 194.

BDS strategies operate across three main domains. Boycotts target Israeli and international companies, as well as cultural, academic, and sporting institutions complicit in rights violations. Divestment campaigns urge banks, pension funds, and universities to withdraw investments from entities linked to Israeli state policies. Sanctions campaigns pressure governments to restrict trade, end military cooperation, and suspend Israel's participation in international bodies.

Supporters frame BDS as an anti-racist human rights initiative inspired by the global movement against apartheid in South Africa, while critics often portray it as politically motivated or discriminatory. The movement's visibility has provoked countermeasures, including anti-BDS legislation in the United States and the United Kingdom, which restricts the ability of public institutions to engage in boycott actions.

BDS illustrates how non-state actors employ sanctioning practices to exert influence on global politics. It also demonstrates how campaigns rooted in civil society can connect economic measures with symbolic and cultural forms of pressure.

Source: Ethical Consumer (2025)

Summary and conclusion

This chapter has positioned sanctions as one of the defining instruments of contemporary geopolitics, while showing how their reach extends well beyond diplomacy and trade into the everyday conditions that shape tourism. Sanctions are not temporary shocks but sustained disruptions that alter mobility, investment, reputation, and accessibility in ways that deeply affect destinations, businesses, and communities. Unlike crises with identifiable beginnings and recoveries, sanctions are open-ended and politically charged, requiring destinations to operate under prolonged constraint rather than to simply "bounce back."

The chapter clarified what counts as a sanction, noting the diversity of measures that span economic, financial, mobility, cultural, and symbolic domains. It examined the historical development of sanctions from early trade exclusions to modern targeted and financialized instruments, highlighting their transformation into routine features of statecraft. It also examined the scales at which sanctions operate, from multilateral regimes authorized by the United Nations to unilateral measures by powerful states and campaigns led by non-state actors. These variations matter greatly for tourism, since they determine how sanctions affect flows of people, capital, and cultural exchange. Across these histories and forms, sanctions emerge as systemic forces that reshape the global conditions of tourism. They influence who can travel, how businesses operate, how destinations are perceived, and what coping strategies communities employ to survive. Yet the resilience that arises under sanctions is uneven, contested, and politically charged, often deepening inequalities even as it sustains livelihoods.

The analysis presented here highlight the need to treat sanctions not as peripheral disruptions but as distinct forms of crisis within tourism studies. They expose the entanglement of tourism with global power relations and challenge conventional assumptions about recovery and resilience. The next chapter turns to tourism itself, situating it explicitly within geopolitical debates. It examines how tourism functions as a geopolitical practice and how sanctions, as instruments of coercion and control, reshape the politics of borders, mobility, and international relations.

Questions for discussion

1. How do sanctions compare with other types of tourism crises in terms of duration, severity, and political complexity?
2. Should sanctions be conceptualized as external "shocks" or as ongoing conditions of governance?
3. What methodological challenges arise when studying the tourism impacts of sanctions compared to natural hazards or pandemics?
4. How might resilience under sanctions look different when viewed from the perspective of states, businesses, or households?
5. In what ways do sanctions reveal the political nature of tourism more clearly than other crises?

References

BDS Movement. (2025). *20 years of the BDS movement.* https://bdsmovement.net/.

Beirman, D. (2020). *Restoring tourism destinations in crisis: A strategic marketing approach.* Routledge.

Brzozowski, A. (2025). *EU countries eye curbs on Russian tourist visas and diplomats.* https://www.euractiv.com/news/eu-countries-mull-restrictions-for-russia-tourists-visas-curbs-on-diplomat-freedom/.

Cortright, D., & Lopez, G. A. (1999). Are sanctions just? The problematic case of Iraq. *Journal of International Affairs*, 735–755.

Cortright, D., & Lopez, G. A. (2018). *Economic sanctions: Panacea or peacebuilding in a post-cold war world?* Routledge.

Crawford, N. C. (1999). Trump card or theater? An introduction to two sanctions debates. In N. C. Crawford & A. Klotz (Eds.), *How sanctions work: Lessons from South Africa* (pp. 3–24). MacMillan.

Ethical Consumer. (2025). *What is the boycott, divestment and sanctions movement?* https://www.ethicalconsumer.org/ethicalcampaigns/boycotts/bds-israel-boycott.

Euractiv (2025). *US, Ukraine and EU officials make progress on updated peace proposal.* https://www.euractiv.com/news/eu-fights-for-seat-at-the-table-as-us-ukraine-talks-advance/.

European Commission. (2025). *Overview of sanctions and related resources.* https://finance.ec.europa.eu/eu-and-world/sanctions-restrictive-measures/overview-sanctions-and-related-resources_en.

Farrall, J. M. (2007). *United Nations sanctions and the rule of law.* Cambridge University Press.

Hu, X. (2016). Legacy of Tiananmen Square incident in Sino-US relations (post-2000). *East Asia*, *33*(3), 213–232.

Hufbauer, G. C. (1998). Economic sanctions. *Proceedings of the ASIL Annual Meeting*, *92*, 332–335.

Hufbauer, G. C., Schott, J. J., Elliott, K. A., & Oeggm, B. (2008). *Economic sanctions reconsidered* (3rd ed.). The Peterson Institute for International Economics.

Jenkins, S. (2025). Trump's tariffs are ruinous, but are western sanctions any better? Just look at Syria. *The Guardian*. https://www.theguardian.com/commentisfree/2025/apr/11/trump-tariffs-western-sanctions-syria-uk.

Jentleson, B. W. (2022). *Sanctions: What everyone needs to know.* Oxford University Press.

Khodadadi, M. (2025). Tourism resilience in Iran: Navigating sanctions, diplomacy and emerging opportunities. *Research in Hospitality Management*, *15*(2), 102–108.

Mair, J., Ritchie, B. W., & Walters, G. (2016). Towards a research agenda for post-disaster and post-crisis recovery strategies for tourist destinations: A narrative review. *Current Issues In Tourism*, *19*(1), 1–26.

Malone, D. M. (2006). *The international struggle over Iraq: Politics in the UN security council 1980–2005.* Oxford University Press.

Moiseienko, A. (2021). Due process and unilateral targeted sanctions. In *Research handbook on unilateral and extraterritorial sanctions* (pp. 405–423). Edward Elgar Publishing.

Mulder, N. (2022). *The economic weapon: The rise of sanctions as a tool of modern war.* Yale University Press.

Nixon, R. (2016). *Selling apartheid: South Africa's global propaganda war.* Pluto Books.

Office of Foreign Assets Control [OFAC]. (2025). *Sanctions programs and country information.* https://ofac.treasury.gov/sanctions-programs-and-country-information.

Pala, M. (2025). *EU considers full ban on Russian tourists in new sanctions package.* https://www.aa.com.tr/en/europe/eu-considers-full-ban-on-russian-tourists-in-new-sanctions-package/3689590.

Pratt, S. A., & Alizadeh, V. (2018). The economic impact of the lifting of sanctions on tourism in Iran: A computable general equilibrium analysis. *Current Issues in Tourism*, *21*(11), 1221–1238.

Rodríguez, F. (2024). The human consequences of economic sanctions. *Journal of Economic Studies*, *51*(4), 942–963.

Rossignol, M. (1996). *Sanctions, the economic weapon in the new world order* (Rev. edn.). Library of Parliament BP-346E.: Research Branch, Political and Social Affairs Division, Library of Parliament, Government of Canada.

Seyfi, S., & Hall, C. M. (2019). International sanctions, tourism destinations and resistive economy. *Journal of Policy Research in Tourism, Leisure and Events*, *11*(1), 159–169.

Seyfi, S., & Hall, C. M. (2020a). *Tourism, sanctions and boycotts*. Routledge.

Seyfi, S., & Hall, C. M. (2020b). Sanctions and tourism: Effects, complexities and research. *Tourism Geographies*, *22*(4–5), 749–767.

Seyfi, S., & Hall, C. M. (2020c). Sanctions and tourism: Conceptualisation and implications for destination marketing and management. *Journal of Destination Marketing & Management*, *15*, 100381.

Seyfi, S., Kimbu, A. N., Tavangar, M., Vo-Thanh, T., & Zaman, M. (2025). Surviving crisis: Building tourism entrepreneurial resilience as a woman in a sanctions-ravaged destination. *Tourism Management*, *106*, 105025.

United Nations. (2025). *United Nations charter, chapter VII: Action with respect to threats to the peace, breaches of the peace, and acts of aggression*. https://www.un.org/en/about-us/un-charter/chapter-7.

van Bergeijk, P. A. (Ed.). (2021). *Research handbook on economic sanctions*. Edward Elgar Publishing.

Wile, R. (2025). As international tourists pull back on U.S. travel and purchases, $90 billion in lost revenue looms. *NBC News*. https://www.nbcnews.com/business/consumer/travelers-going-to-united-states-less-due-to-trump-tariffs-rhetoric-rcna201640.

York, E., & Durante, A. (2025). Trump tariffs: The economic impact of the trump trade war. *Tax Foundation*. https://taxfoundation.org/research/all/federal/trump-tariffs-trade-war/.

Zhang, H., & Zhu, H. (2024). Investigating the transformative power of brand love in destination crisis recovery. *Journal of Travel Research*, 00472875241296086.

2 Tourism, mobility, and the geopolitics of sanctions

Introduction

As outlined in the previous chapter, sanctions have become one of the most widely used tools of international politics. Multilateral organizations, regional bodies, and powerful states increasingly rely on them as alternatives to military intervention and as instruments of foreign policy (van Bergeijk, 2021; Mulder, 2022). While their geopolitical purpose is often framed in terms of deterring aggression, enforcing international law, or expressing political condemnation, sanctions inevitably spill over into economic, social, and cultural domains (Hufbauer et al., 2008). Among the sectors directly affected is tourism, which depends on the free flow of people, capital, and images across borders (Khalid et al., 2024). Despite the importance of these linkages, the relationship between sanctions and tourism remains underexplored only until recently. While a substantial body of work in political science, international relations, and economics has examined the motivations behind sanctions, their effectiveness, and their impact on targeted economies (e.g., Cortright & Lopez, 1999, 2018; Hufbauer et al., 2008; Oghazi et al., 2024), tourism has only recently attracted sustained attention in this context.

To understand how sanctions shape tourism, this chapter positions tourism within its broader geopolitical context. Building on the previous discussion of sanctions as instruments of statecraft, it argues that tourism is not only an economic or cultural activity but also a political practice embedded in shifting relations of power. The chapter begins by examining how tourism has increasingly been conceptualized as geopolitical, with borders, mobility, and statecraft shaping both the possibilities and limits of leisure and travel. It then considers sanctions as deliberate geopolitical interventions that restrict movement, reconfigure cross-border relations, and stigmatize destinations. Framing sanctions as geopolitical crises highlights how global politics translate into everyday disruptions in tourism, exposing the sector's deep entanglement with international struggles over authority, legitimacy, and representation.

DOI: 10.4324/9781003654339-2

Tourism as a geopolitical practice

Tourism has increasingly been recognized as a geopolitical practice rather than a phenomenon detached from politics. Early contributions hinted at this link, with Richter (1989) and Hall (1994) arguing that tourism was inseparable from state political agendas. At the time, their work received limited uptake, partly because tourism was still framed primarily as a cultural or economic activity. Yet as international travel expanded in scale and significance, the political dimensions of tourism became more difficult to ignore. This shift in attention parallels broader debates in geopolitics itself.

Traditionally, geopolitics referred to the influence of geography on political practice and discourse, with a strong emphasis on international competition between nation states (Dodds, 2007). Since the 1990s, however, the critical turn in geopolitics has broadened the term's connotations to include material and affective practices, everyday experiences, and situated encounters (Ó Tuathail & Dalby, 1998; Gillen & Mostafanezhad, 2019). In this expanded sense, geopolitics is not only about territory and borders but also about how cultural discourses and texts coproduce geopolitical imaginations and their material consequences. Tourism participates directly in these processes by shaping how the world is represented, narrated, and consumed.

This recognition has inspired what Mostafanezhad (2018) termed the "geopolitical turn" in tourism geographies, which has brought sustained attention to the ways in which travel and leisure are implicated in global power struggles. Tourism is not only about leisure but is as much a political terrain as a cultural practice (Molz, 2010). From this perspective, tourism contributes to geopolitics both through the visible movement of people across borders and through the symbolic production of places as desirable, dangerous, contested, or inaccessible (Mostafanezhad & Norum, 2016). State travel advisories, media portrayals, and destination marketing campaigns all participate in this symbolic ordering, mapping out global risks and security threats in ways that position destinations within political hierarchies (Bianchi, 2006).

The growing body of literature illustrates that tourism and geopolitics are not separate realms but mutually constitutive. Azcárate (2024) and Huang and Liu (2024) emphasize that the two are woven together in everyday practices of leisure, diplomacy, and infrastructure development. Azcárate et al. (2021) describe tourism geopolitics as the ways in which power is "wielded, bent, or suffered in and across geographical scales" through tourism practices (p. 4). This resonates with Gillen's (2025) observation that the concept is "trending, buzzing, and subject to fascinating studies" (p. 632). Gillen highlights that tourism geopolitics combines rootedness with mobility, linking everyday leisure practices to broader processes of bordering, diplomacy, and infrastructural development. Such perspectives highlight tourism's role as one of the most far-reaching geopolitical practices of the twenty-first century (Mostafanezhad, 2018; Hall, 2017).

The political stakes of tourism become particularly visible during crises. Wars, terrorism, pandemics, and sudden border closures demonstrate the dependence of international tourism on stable political conditions (Rowen, 2016; Huang, 2022). The COVID-19 pandemic, for instance, made clear how quickly political decisions over borders and mobility could paralyze global travel. Yet the entanglement extends beyond moments of disruption (Seyfi et al., 2023b). The globalization of tourism supply chains, the rise of international tourism corporations, and the sheer volume of cross-border flows further entrench tourism as an arena in which governments assert ideologies, exercise power, and pursue foreign policy objectives (Mostafanezhad, 2018; Mostafanezhad et al., 2021; Iaquinto et al., 2024).

Nevertheless, a recurring challenge in the literature is the disciplinary divide. Tourism has often been associated with leisure, consumption, and economic discourse, while geopolitics has centered on violence, territoriality, and power relations (Agnew, 2004; Dodds, 2007; Franklin & Crang, 2001; Tribe, 2008). As Hall (2017) observed, the two fields have often been treated as mutually exclusive because they rest on "conflicting interpretations, contrasting methodologies, and diverse theorisations" (p. 3). Contemporary scholarship, however, insists that they must be brought together, since the spatial consequences of international relations are evident not only at the supranational and national levels but also in local destinations and at international borders (Mostafanezhad, 2018; Mostafanezhad et al., 2021; Azcárate, 2024; Iaquinto et al., 2024) (Box 2.1).

Box 2.1 Geopolitical encounters of tourism

Gillen and Mostafanezhad argue that the ordinary *tourism encounter*, the routine, embodied meeting of tourists, residents, places, objects, and meanings, is itself geopolitical. Rather than reserving geopolitics for statecraft and summitry, they relocate it in the mundane frictions, performances, and representations that constitute travel. Encounters are place-based and everyday, yet they reverberate across scales: what happens at a resort gate, a visa counter, a heritage site, or on a volunteer project loops into national narratives, regional rivalries, and global imaginaries (Gillen & Mostafanezhad, 2019). At the heart of their framework is the claim that tourism both reflects and *produces* geopolitics. It does so through discourse (destination branding, travel advisories, news and social media) and through practice (itineraries, rules, etiquettes, infrastructures) that include and exclude, normalize and contest. This is "everyday geopolitics" in action: visa queues, airport screenings, guided tours, humanitarian selfies, and review platforms operate as seemingly banal sites where power is enacted, borders are felt, and identities are negotiated. As these micro-events circulate through media and bureaucracy, they harden into policies, advisories, blacklists, etiquette campaigns, thereby stitching intimate encounters to formal instruments of governance.

They organize these processes around three overlapping modes of encounter. *Temporal encounters* highlight heritage and memory: decisions about what to conserve and how to stage it bind pasts to presents and futures, mobilizing nostalgia and aspiration to legitimize authority, territorial claims, and development pathways. *Bodily encounters* highlight the sensing, feeling body as a medium of power. Volunteer tourism, for example, wraps geopolitical hierarchies in intimate gestures of care; acts of "help" can reproduce North–South inequalities, advance soft power, and naturalize uneven development, even as participants report transformative experiences. *Identity encounters* bring the self/other relation to the surface. Host–guest exchanges and group tour scripts crystallize or unsettle categories such as citizen/foreigner or same/different nation, as seen in Cross-Strait tourism where itineraries and talk can either soften or harden claims of cultural sameness and sovereignty (Gillen & Mostafanezhad, 2019).

Media circuits amplify these encounters. Viral stories of "misbehaving tourists," humanitarian imagery, and influencer narratives discipline bodies and codify global conduct, while also offering counter-scripts that critique saviorism or expose inequality. Communities, for their part, are not passive recipients of touristic scripts: they accept, reshape, or refuse them, sometimes "refusing to be toured," thereby redirecting heritage temporalities and unsettling dominant development logics. In this way, micro-decisions at the destination scale can elevate outward, becoming geopolitical events in their own right.

This conceptual lens matters because it bridges critical geopolitics (everyday, embodied, affective) with tourism studies (encounter, performance, mobilities), expanding what counts as "impact." Geopolitical meaning is not only in arrivals and receipts but in the practices that normalize borders, securitize spaces, perform nationhood, and animate resistance. It also opens fertile research pathways: emotions and affect in geopolitical tourism; the political life of images; the (post) colonial intimacies of volunteering; race, class, and gender in encounter; human–nonhuman assemblages in conservation tourism; and the researcher's own positionality in representing these interactions.

The takeaway is succinct: encounters are the mundane machinery through which tourism *wields*, *bends*, or *suffers* power. Reading tourism through temporal, bodily, and identity encounters shows how global politics is stitched from ordinary meetings in ostensibly apolitical spaces of leisure, and why those spaces matter for nation-building, development, territorialization, and resistance (Gillen & Mostafanezhad, 2019).

Source: After Gillen & Mostafanezhad (2019)

Borders, mobility, and the geopolitics of sanctions

As previously discussed, tourism and geopolitics are deeply entangled. This relationship becomes most visible in the sphere of borders and mobility, where the regulation of movement functions not only as a practical condition for travel but also as a strategic instrument of political control. Mobilities, borders, and geopolitics are increasingly recognized as inseparable domains (Cresswell, 2010; Rowen, 2018). As Rumford (2006, p. 155) observed, "to theorize mobilities and networks is at the same time to theorize borders." Tourists, as particular kinds of mobile subjects, navigate a bordered world in which their movements both shape and are shaped by the construction and performance of borders (Rowen, 2018).

The border has long been central to geopolitical studies, which examine how states materialize and perform distinctions between the "domestic" and the "foreign." Through globalization, borders have become more open to flows of people and capital yet also sites of renewed contestation (Qu, 2025).

Two key strands emerge from this analysis: border securitization and cross-border reterritorialization. Tourism is not simply subject to these dynamics but is often utilized by states as a tool of authority, territorialization, and reterritorialization (Hazbun, 2004; Rowen, 2017). The Chinese government's mobilization of tourism to consolidate claims over Taiwan, Xinjiang, and the South China Sea illustrates how states use tourism strategically to reinforce sovereignty (Huang & Suliman, 2023; Szadziewski et al., 2022) (Box 2.2).

At the same time, international tourism mobilities have expanded dramatically across time and space. Hall (2015) cautioned that while overall growth is significant, the dominance of North American and Eurocentric perspectives requires scrutiny. Greater attention to emerging economies, which increasingly shape both demand and supply in global tourism, provides a fuller picture. Hall (2005) further linked travel expansion to demographic and cultural shifts such as population growth, urbanization, and consumer cultures that prize mobility. Yet tourism mobilities remain highly contingent on political relations between states (Hall, 2017). Visa regimes, bilateral agreements, and airspace rights depend on diplomatic stability. When cooperation prevails, policies such as visa waivers, open skies agreements, and investment incentives can ease flows. The Schengen Area illustrates how integration reduces

Box 2.2 Tourism sanctions as a means of economic statecraft

Using fuzzy-set Qualitative Comparative Analysis (fsQCA) across 18 positive/negative cases, the authors identify *configurations* under which China utilize outbound tourism restrictions as coercive diplomacy. They find tourism sanctions are most likely when hard security disputes are present and either (i) the target is highly dependent on Chinese visitors but *not* on tourism overall, or (ii) the target is a tourism-dependent democracy. In practice, China operationalizes control via ADS status, licensing of outbound agencies, verbal/administrative directives, flight/charter approvals, and coordinated media/travel warnings. Illustrative episodes include THAAD-related curbs on South Korea (sharp collapse of group tours), softer, indirect limits toward Japan during the 2012 Senkaku/Diaoyu dispute, suspensions toward the Philippines amid maritime tensions, stepwise reductions to Taiwan linked to sovereignty politics, targeted pressure on Palau over its Taiwan ties, and episodic bans on North Korea after nuclear tests. The study argues tourism sanctions typically *signal* and precede broader packages rather than aim to crash an entire economy; their logic is a focused "soft punch" calibrated to create sectoral pain and political salience. Two enabling conditions appear necessary: a large, steadily growing outbound market and strong state control over its organization (e.g., ADS, licensing). Policy implications emphasize targets' pivots to FIT markets, visa facilitation, and niche/loyalty strategies to cushion sudden group-tour withdrawals.

Source: Chung et al. (2025)

bureaucratic barriers and fosters tourism. Conversely, strained relations can harden borders abruptly, as in the deterioration of ties between China and South Korea following the 2017 THAAD missile deployment, which led to a ban on Chinese group tours and significant losses for South Korea's inbound tourism (BBC, 2017) (Box 2.3).

Air connectivity is another site of vulnerability. Overflight rights, negotiated diplomatically, underpin the global airline network, and their suspension can reshape routes and increase costs. The closure of European, Canadian, and U.S. airspace to Russian carriers in 2022, and the reciprocal bans imposed by Russia, forced airlines to reroute long-haul flights between Europe and Asia (Xue et al., 2025). Similarly, the Gulf blockade of Qatar in 2017 pressured Qatar Airways to rely on Iranian and Turkish airspace, raising operational costs and complicating passenger flows (Wintour, 2017).

Box 2.3 South Korea tourism hit by China ban

In 2017, South Korea's decision to host the United States' Terminal High Altitude Area Defense (THAAD) system triggered a sharp downturn in its tourism industry. In response, China imposed restrictions on outbound group tours to South Korea, a move widely interpreted as an informal sanction designed to signal Beijing's disapproval. The effects were immediate: Chinese arrivals, which had accounted for nearly half of South Korea's inbound market, collapsed within months, producing billions of dollars in estimated losses for airlines, hotels, and retailers. The episode illustrates how geopolitical disputes can quickly reshape mobility flows, with borders that had appeared open to mass tourism suddenly hardened by political retaliation. It also highlights the vulnerability of destinations heavily dependent on a single source market. In South Korea's case, diversification strategies were accelerated in the aftermath of the dispute, with authorities seeking to attract visitors from Southeast Asia, Japan, and Western markets. Yet the THAAD episode demonstrated the precarious position of tourism in international politics, where shifts in diplomatic relations can reverberate through everyday practices of leisure and consumption.

Source: BBC (2017)

Visa and entry restrictions serve as perhaps the clearest example of borders functioning as geopolitical tools. Beyond regulating mobility, they signal alliances, rivalries, and identities. The U.S. travel bans of 2017, which targeted citizens of several Muslim-majority states, reduced tourist arrivals and disrupted family visits, student exchanges, and professional travel while sending strong political messages about national security and identity (Arafa, 2018) (Box 2.4). Such measures extend their impact beyond enforcement by shaping global perceptions and stigmatizing entire populations. These practices invite wider debates over mobility rights. Freedom of movement is often described as a human rights principle, codified in Article 13 of the Universal Declaration of Human Rights, which affirms the right to leave and return to one's country (Gilbert, 2014). Yet in practice, no automatic right exists to enter another state. Citizens may be legally entitled to depart but still denied visas for their destinations. Even regional integration produces only conditional rights. As Coles (2008, p. 67) noted of EU accession in 2004, "it is all well and good to be offered the right to travel but this is meaningless if the right to entry is restricted."

Legal restrictions provide the foundation for sanctions targeting cross-border movement. Yet the legal cannot be disentangled from political and

Box 2.4 The U.S. travel ban

In June 2025, President Donald Trump signed an executive order imposing a sweeping new travel ban on nationals of twelve countries, including Afghanistan, Iran, Libya, Somalia, Sudan, and Yemen. Citizens of seven additional states, among them Cuba and Venezuela, faced partial restrictions. The White House justified the ban as a national security measure, arguing that the targeted countries had failed to meet "adequate vetting standards."

Although exceptions were provided for lawful permanent residents, certain visa holders, and participants in major sporting events, the ban drew immediate criticism for its scope and discriminatory effects. The African Union condemned the measures, while Chad retaliated by suspending visas for U.S. citizens. Domestic opponents described the order as an expansion of Trump's earlier "Muslim ban," warning that it would isolate the United States diplomatically.

For tourism, the consequences were twofold. On one hand, citizens of banned countries lost direct access to U.S. destinations, blocking mobility for leisure, business, and family visits. On the other, the ban signaled to international audiences that the United States was becoming a more exclusionary destination, potentially damaging its global image as an open society. Travel restrictions of this kind demonstrate how geopolitical decisions reverberate beyond security, shaping perceptions of accessibility, discouraging flows, and producing uncertainty for airlines, tour operators, and travelers alike.

Source: BBC News (2025)

social dimensions. Visa blacklisting, travel bans, and entry denials are closely tied to xenophobia, racism, and Islamophobia, reinforcing forms of discrimination that deepen exclusions (Torabian & Miller, 2017). Heightened security regimes shape not only who is permitted to move but also the conditions under which mobility takes place, embedding inequality and stigma into the very structures of travel. Findings from Torabian and Miller (2017) demonstrate how racialized identities and lived experiences at borders produce frustration, humiliation, and uneven access to the right to travel, highlighting that the much-celebrated "freedom of movement" is far from universal. More recently, Torabian and Muldoon (2024) show how vaccine passports, while framed as tools of reopening and safety, risk reproducing and elevating these exclusions by privileging citizens of wealthy states and further immobilizing those in the global periphery. Parallel insights are provided by Seyfi et al. (2023b), who argue that international travel restrictions during the COVID-19 pandemic revealed both the selectivity of border controls and the disproportionate burdens imposed on sanctioned states, where health crises were exacerbated by the continuation of coercive foreign policy.

Sanctions as geopolitical crises

Sanctions differ from other tourism crises because they are not accidental shocks such as natural disasters, pandemics, or terrorist attacks. Instead, they are deliberately imposed instruments of statecraft, designed to constrain political adversaries through economic, financial, and mobility restrictions (Seyfi & Hall, 2019, 2020a, 2020b). This external and intentional character sets sanctions apart, turning them into geopolitical crises that emerge slowly, last for extended periods, and reshape both material and symbolic conditions for tourism.

Tourism is directly implicated in these crises because sanctions strike at the foundations of international mobility (Seyfi et al., 2022a). Travel bans, restrictions on airspace, prohibitions on financial transactions, and the exclusion of sanctioned states from global booking and payment systems all destabilize the sector. For example, the sanctions imposed on Iran effectively isolated the country from the international financial system, collapsing credit card services and forcing both tourists and operators to rely on cash-based transactions. Although the 2015 Joint Comprehensive Plan of Action temporarily lifted restrictions and led to a surge in arrivals, the reimposition of U.S. sanctions in 2018 triggered a rapid withdrawal of international firms and the abandonment of planned investments (Khodadadi, 2016, 2018).

The use of sanctions to restrict movement demonstrates most directly how the geopolitics of borders is entangled with foreign policy. Long-standing U.S. restrictions on travel to Cuba demonstrate efforts to constrain mobility as part of a broader strategy to influence political reform and weaken regional alliances (Sullivan, 2011). Research on Cuba's air transport network shows that the embargo not only curtailed direct connectivity with the United States for more than five decades but also forced Cuba to reconfigure its aviation system by seeking alternative alliances and diversifying its international reach (Córdoba, 2022). Similarly, the 2017 executive order signed by President Trump suspended entry for citizens of several Muslim-majority states, transforming border control into a weapon of statecraft (Arafa, 2018). These measures illustrate how mobility regulation becomes a coercive instrument, signaling political divides and shaping international behavior. Tourism, therefore, sits at the heart of this struggle over mobility. The sector depends on openness, accessibility, and connectivity, yet these conditions are continuously reshaped by political negotiations, alliances, and conflicts. Sanctions make this dependence explicit: by restricting visas, closing airspaces, and stigmatizing destinations, they show how borders and mobility can be deliberately weaponized. What is leisure for some becomes, for others, a stark reminder that the right to travel is contingent, conditional, and profoundly geopolitical. Seen in this way, sanctions constitute a distinct category of geopolitical crisis. They externalize global conflicts into the everyday functioning of tourism systems, transforming borders into weapons, mobility into leverage, and destinations into symbols of contestation. Their effects are not confined to the sanctioned states themselves but ripple across regions and industries, reinforcing the need to situate tourism research firmly within the geopolitical analysis of sanctions.

Summary and conclusion

This chapter has argued that tourism must be understood as a geopolitical practice, inseparable from the shifting configurations of power, borders, and mobility. Building on early scholarship that linked tourism to state agendas, the discussion has examined how critical geopolitics expanded the field to include everyday practices, symbolic discourses, and infrastructural arrangements that shape how destinations are imagined and accessed. Tourism thus emerges not only as a cultural and economic phenomenon but as a site where authority, legitimacy, and representation are continually negotiated.

Particular attention was given to borders and mobility as central arenas where geopolitics and tourism converge. Visa regimes, travel bans, airspace restrictions, and cross-border agreements demonstrate how the regulation of movement functions as both a practical condition for tourism and a strategic tool of foreign policy. Empirical examples, from the deterioration of China–South Korea relations during the THAAD dispute to the restrictions imposed on Russian carriers after 2022, illustrate how geopolitical struggles materialize in tourism systems, often with sudden and destabilizing effects. Scholarship on racialized and unequal experiences of mobility further underscores that these processes are not neutral but embed hierarchies and exclusions into the very structures of international travel.

Against this backdrop, sanctions can be understood as a distinct form of geopolitical crisis. Unlike natural disasters or pandemics, they are deliberately imposed, prolonged, and politically charged instruments of coercion. Their impacts extend across multiple domains of tourism, from visa regimes and air connectivity to financial transactions and destination reputations. Case examples such as the long-standing U.S. embargo on Cuba or the U.S. travel bans on Muslim-majority countries show how sanctions weaponize borders and mobility, transforming tourism from a domain of leisure into an arena of political contestation. For some destinations, sanctions force adaptation through new alliances and domestic markets; for others, they deepen isolation, precarity, and exclusion. Sanctions, therefore, expose tourism's deep entanglement with geopolitics, not as a peripheral effect but as a structural condition. They demonstrate how global conflicts and power struggles reverberate through the infrastructures, practices, and imaginaries of tourism.

The next chapter builds on this foundation by examining in greater detail the consequences of sanctions for tourism destinations, exploring how these measures disrupt flows, reshape competitiveness, and compel adaptation across different contexts.

Questions for discussion

1. To what extent can tourism be considered a geopolitical practice in its own right, rather than a secondary effect of political processes?
2. How do borders, visa regimes, and travel bans illustrate the use of mobility as an instrument of statecraft?
3. In what ways do sanctions differ from other tourism-related crises such as pandemics or natural disasters?
4. How do racialized, gendered, or classed experiences of border control complicate the idea of universal "freedom of movement"?
5. What are the ethical implications of using mobility restrictions as a coercive tool, particularly when they disproportionately affect ordinary travelers and workers?
6. How might tourism research contribute to wider debates in international relations about the legitimacy and effectiveness of sanctions?

References

Agnew, J. (2004). *Geopolitics: Re-visioning world politics*. Routledge.

Arafa, M. (2018). A question to the president of the United States, Donald Trump: Is it a travel ban, or a Muslim ban, or a travel Muslim ban? *Revista de Investigações Constitucionais*, *5*(2), 9–33.

Azcárate, M. (2024). Tourism geopolitics: Routes and worldings. *Tourism Geographies*, *26*(5), 814–832.

Azcárate, M. C., Mostafanezhad, M., & Norum, R. (2021). Introduction: Tourism geopolitics. In M. Mostafanezhad, M.C. Azácarte, and R. Norum (Eds.), *Tourism geopolitics: Assemblages of infrastructure, affect, and imagination* (pp. 1–28). University of Arizona Press.

BBC. (2017). *South Korea tourism hit by China ban*. https://www.bbc.com/news/business-40565119.

BBC News. (2025). *Trump travel ban barring citizens from 12 countries takes effect*. https://www.bbc.com/news/articles/cz9y72v5yv5o.

Bianchi, R. (2006). Tourism and the globalisation of fear: Analysing the politics of risk and (in) security in global travel. *Tourism and Hospitality Research*, *7*(1), 64–74.

Chung, J. Y., Paik, W., & Kim, Y. H. (2025). Tourism sanctions as a means of China's economic statecraft: An fsQCA analysis. *Anatolia*, *36*(2), 395–407.

Coles, T. (2008). Citizenship and the state: Hidden features in the internationalisation of tourism. In T. Coles & C. M. Hall (Eds.), *International business and tourism* (pp. 69–83). Routledge.

Córdoba, H. A. (2022). The evolution of Cuba's air transport network in the Embargo Era: 1967–2017. *Journal of Transport Geography*, *98*, 103259.

Cortright, D., & Lopez, G. A. (1999). Are sanctions just? The problematic case of Iraq. *Journal of International Affairs*, 735–755.

Cortright, D., & Lopez, G. A. (2018). *Economic sanctions: Panacea or peace-building in a post-cold war world?* Routledge.

Cresswell, T. (2010). Towards a politics of mobility. *Environment and Planning D: Society and Space*, *28*(1), 17–31.

Dodds, K. (2007). *Geopolitics: A very short introduction*. Oxford University Press.

Franklin, A., & Crang, M. (2001). The trouble with tourism and travel theory? *Tourist Studies*, *1*(1), 5–22.

Gilbert, J. (2014). *Nomadic peoples and human rights*. Routledge.

Gillen, J. (2025). Tourism geopolitics: Roots and branches. *Tourism Geographies*, *27*(3–4), 631–641.

Gillen, J., & Mostafanezhad, M. (2019). Geopolitical encounters of tourism: A conceptual approach. *Annals of Tourism Research*, *75*, 70–78.

Hall, C. M. (1994). *Tourism and politics: Policy, power and place*. Wiley.

Hall, C. M. (2005). *Tourism: Rethinking the social science of mobility*. Pearson Education.

Hall, C. M. (2015). On the mobility of tourism mobilities. *Current Issues in Tourism*, *18*(1), 7–10.

Hall, D. R. (2017). Tourism and geopolitics: Issues and concepts from central and Eastern Europe. *CABI*.

Hazbun, W. (2004). Globalisation, reterritorialisation and the political economy of tourism development in the middle east. *Geopolitics*, *9*(2), 310–341.

Huang, Y. (2022). Consuming geopolitics and feeling maritime territoriality: The case of China's patriotic tourism in the South China Sea. *Political Geography*, *98*, 102669.

Huang, Y., & Liu, Y. (2024). A critical reflection on tourism geopolitics: Research progress and future agenda. *Tourism Geographies*, *26*(5), 881–897.

Huang, Y., & Suliman, S. (2023). Geopolitics,(re) territorialisation, and China's patriotic tourism in the South China Sea. *Geopolitics*, *28*(4), 1465–1488.

Hufbauer, G. C., Schott, J. J., Elliott, K. A., & Oeggm, B. (2008). *Economic sanctions reconsidered* (3rd ed.). The Peterson Institute for International Economics.

Iaquinto, B. L., Gillen, J., & Mostafanezhad, M. (2024). The new tourism geopolitics: Bridging tourism geographies and political geography. *Tourism Geographies*, *26*(5), 723–734.

Khalid, U., Okafor, L., & Burzynska, K. (2024). Sanctions and tourist flows: The roles of religion and geography. *Annals of Tourism Research Empirical Insights*, *5*(2), 100143.

Khodadadi, M. (2016). A new dawn? The Iran nuclear deal and the future of the Iranian tourism industry. *Tourism Management Perspectives*, *18*, 6–9.

Khodadadi, M. (2018). Donald Trump, US foreign policy and potential impacts on Iran's tourism industry: Post-nuclear deal. *Tourism Management Perspectives*, *26*, 28–30.

Molz, J. G. (2010). Performing global geographies: Time, space, place and pace in narratives of round-the-world travel. *Tourism Geographies*, *12*(3), 329–348.

Mostafanezhad, M. (2018). The geopolitical turn in tourism geographies. *Tourism Geographies*, *20*(2), 343–346.

Mostafanezhad, M., Azcárate, M. C., & Norum, R. (2021). *Tourism geopolitics: Assemblages of infrastructure, affect, and imagination*. University of Arizona Press.

Mostafanezhad, M., & Norum, R. (2016). Towards a geopolitics of tourism. *Annals of Tourism Research*, *61*(1), 226–228.

Mulder, N. (2022). *The economic weapon: The rise of sanctions as a tool of modern war*. Yale University Press.

Oghazi, P., Patel, P. C., & Hajighasemi, A. (2024). Gendered crisis approach: Exploring the gendered impact of Iranian sanctions on nascent entrepreneurship outcomes. *International Small Business Journal*, *42*(8), 1016–1046.

Ó Tuathail, G., & Dalby, S. (Eds.). (1998). *Rethinking geopolitics*. Routledge.

Qu, Y. (2025). Tourism development on China's Northwestern border: Territorialisation, reterritorialisation, and geopolitics of the great altai region in inner Asia. *Geopolitics*. https://doi.org/10.1080/14650045.2025.2462955.

Richter, L. K. (1989). *The politics of tourism in Asia*. University of Hawaii Press.

Rowen, I. (2016). The geopolitics of tourism: Mobilities, territory, and protest in China, Taiwan, and Hong Kong. *Annals of the American Association of Geographers*, *106*(2), 1–9.

Rowen, I. (2017). Touring in heterotopia: Travel, sovereignty, and exceptional spaces in Taiwan and China. *Asian Anthropology*, *16*(1), 20–34.

Rowen, I. (2018). The geopolitics of tourism: Mobilities, territory, and protest in China, Taiwan, and Hong Kong. In *Geographies of mobility* (pp. 143–151). Routledge.

Rumford, C. (2006). Theorizing borders. *European Journal of Social Theory*, *9*(2), 155–169.

Seyfi, S., & Hall, C. M. (2019). International sanctions, tourism destinations and resistive economy. *Journal of Policy Research in Tourism, Leisure and Events*, *11*(1), 159–169.

Seyfi, S., & Hall, C. M. (2020a). *Tourism, sanctions and boycotts*. Routledge.

Seyfi, S., & Hall, C. M. (2020b). Sanctions and tourism: Effects, complexities and research. *Tourism Geographies*, *22*(4–5), 749–767.

Seyfi, S., Hall, C. M., & Shabani, B. (2023b). COVID-19 and international travel restrictions: The geopolitics of health and tourism. *Tourism Geographies*, *25*(1), 357–373.

Seyfi, S., Hall, C. M., & Vo-Thanh, T. (2022a). Tourism, peace and sustainability in sanctions-ridden destinations. *Journal of Sustainable Tourism*, *30*(2–3), 372–391.

Sullivan, M. P. (2011). *Cuba: US restrictions on travel and remittances*. DIANE Publishing.

Szadziewski, H., Mostafanezhad, M., & Murton, G. (2022). Territorialization on tour: The tourist gaze along the Silk Road economic belt in Kashgar, China. *Geoforum*, *128*, 135–147.

Torabian, P., & Miller, M. C. (2017). Freedom of movement for all? Unpacking racialized travel experiences. *Current Issues in Tourism*, *20*(9), 931–945.

Torabian, P., & Muldoon, M. (2024). Covid-19 vaccine passports: Global inequalities and entangled mobilities. *Tourism Culture & Communication*, *24*(2), 145–154.

Tribe, J. (2008). Tourism: A critical business. *Journal of Travel Research*, *46*(3), 245–255.

van Bergeijk, P. A. (Ed.). (2021). *Research handbook on economic sanctions*. Edward Elgar Publishing.

Wintour, P. (2017). Gulf plunged into diplomatic crisis as countries cut ties with Qatar. *The Guardian*. https://www.theguardian.com/world/2017/jun/05/saudi-arabia-and-bahrain-break-diplomatic-ties-with-qatar-over-terrorism.

Xue, D., Du, S., Xu, Y., Zhang, Q., & Sun, X. (2025). Airspace closure challenges: Exploring the impact of the Russia-Ukraine conflict on flight operations and pathways to solutions. *Transportation Research Interdisciplinary Perspectives*, *31*, 101396.

3 Sanctions crisis and tourism destinations

Introduction

The earlier chapters established sanctions as a distinctive geopolitical condition. Unlike natural hazards or sudden crises, sanctions are deliberate, prolonged, and externally imposed measures that reorganize the very foundations of mobility and exchange. They reshape borders, visa systems, and financial infrastructures, while simultaneously redefining which destinations are positioned as open, desirable, or legitimate within the global tourism system. This broader framing sets the stage for the present chapter, which turns from conceptual positioning to the concrete outcomes of sanctions for tourism systems and destinations. Sanctions are not sudden or bounded events but prolonged and externally imposed pressures that reverberate through multiple layers of tourism. They reshape international arrivals, investment flows, aviation and cruise networks, and the reputational standing of destinations. At the same time, their impacts extend into the everyday lives of tourism workers, businesses, and communities, altering livelihoods and intensifying existing vulnerabilities. By exploring these outcomes, the chapter highlights how sanctions transform not only the scale of tourism activity but also its quality, accessibility, and sustainability.

Research on sanctions and tourism: key cases and themes

Before examining specific consequences of sanctions for tourism, it is important to take stock of the existing research that has sought to address this intersection. Early work in this field emphasized a striking gap: while sanctions have been extensively examined in political science, international relations, and economics, tourism was largely absent from these debates until recently (Seyfi & Hall, 2019). Subsequent conceptual work extended this agenda by examining implications for destination marketing and management, including exchange rates, investment flows, and negative reputational spillovers (Seyfi & Hall, 2020c). Macro-level econometric studies have since

DOI: 10.4324/9781003654339-3

demonstrated that sanctions consistently depress international tourist flows. Khalid et al. (2024) employed a global gravity dataset and found that sanctions, whether arms, trade, finance, or travel restrictions, reduced bilateral tourist flows. Religious and geographical proximity could soften, but not eliminate, the decline. Similarly, Chi et al. (2024) showed that international travel sanctions significantly reduced arrivals, with stronger effects when both sender and target were developed economies.

Research at the country level shows how sanctions trigger cycles of disruption and adaptation. Pratt and Alizadeh (2018) used a computable general equilibrium model to estimate the economic effects of lifting sanctions on Iran in 2016, projecting tourism gains but warning against unsustainable expansion. More recently, Khodadadi (2025) emphasized Iran's resilience strategies under renewed U.S. pressure, including pivoting to non-Western partnerships, alternative financial systems, and domestic and religious tourism. Research on second-home tourism in Iran highlights currency devaluation under sanctions, which fueled speculation, informal economies, and corruption in rural housing markets (Torabi et al., 2024).

Russian cases highlight sectoral variations. A survey of 103 hotel managers showed divergent impacts by property size, chain affiliation, and location, with cost-cutting and intensified marketing as common adaptations (Ivanov et al., 2017). Aviation-focused studies found that sanctions imposed after 2022 produced spillovers that hit airlines harder than other aviation-related firms, especially large long-haul carriers, and with uneven impacts across regions (Corbet et al., 2025). Earlier work identified the 2014–2015 collapse in outbound Russian tourism, linked to sanctions and ruble devaluation (Ovcharov et al., 2015).

A parallel strand treats tourism itself as an instrument of statecraft. Chung et al. (2025) showed that outbound tourism bans are used strategically when targets are tourism-dependent democracies or heavily reliant on a specific source market. Such measures are designed for political signaling and limited disruption rather than economic collapse. This confirms that tourism can serve both as an arena of impact and as a deliberate foreign policy tool.

Justice- and rights-oriented perspectives broaden the scope further. Seyfi et al. (2023a) examined sanctions in Iran through a justice lens, finding that they constrained mobility rights, communication rights, and economic rights, undermining tourism's capacity to advance the Sustainable Development Goals. Seyfi et al. (2022b) extended this work to women's empowerment, examining how sanctions eroded economic and social opportunities for women in tourism. Toubes and Araújo-Vila (2021) analyzed travel advisories as covert tools of sanctioning, noting the persistent use of risk-related language that stigmatizes destinations long after conditions stabilize. More recent studies demonstrate how Iranian women entrepreneurs in tourism developed gendered entrepreneurial resilience as a form of resistance to both sanctions and domestic patriarchal constraints (Seyfi et al., 2025; Hosseini et al., 2022). Li and Zhang (2021), focusing on the Chinese border city of Dandong, showed how sanctions on North Korea reverberated locally, damaging cross-border trade and tourism while also shaping residents' perceptions of opportunity and risk.

Methodological reflections add another important dimension. Earl (2021) described how sanctions constrain qualitative research by limiting access to data and restricting knowledge exchange, raising important questions about how geopolitical conditions shape scholarship itself. COVID-19 research has also shown the overlap between sanctions and health crises, with border closures and travel restrictions amplifying the isolation of already sanctioned destinations (Seyfi et al., 2023b). Table 3.1 offers an overview of empirical and conceptual research on the tourism impacts of sanctions.

Table 3.1 Overview of research on the tourism impacts of sanctions

Authors / year	*Type of paper*	*Case / context*	*Key findings*
Seyfi & Hall (2020a)	Scoping review	Global	Identified four types of sanctions (financial, sectoral, diplomatic, individual); tourism affected through mobility restrictions, payment systems, supply chains, and destination image.
Seyfi & Hall (2019)	Conceptual paper	Global	Linked sanctions to destination marketing and management; highlighted impacts on exchange rates, foreign investment, and reputational damage.
Khalid et al. (2024)	Econometric (gravity model)	206 origin, 189 destination countries	Sanctions consistently reduce bilateral tourist flows; religious and geographic proximity soften but do not remove negative effects.
Chi et al. (2024)	Econometric (gravity model)	Global, 1995–2015	Travel sanctions reduce international arrivals; stronger effects between developed countries.
Pratt & Alizadeh (2018)	Computable General Equilibrium model	Iran	Lifting sanctions boosts tourism and economy; risks of resource diversion from other sectors; moderate growth advised.
Seyfi & Hall (2018)	Case study	Iran	Introduced concept of "resistive economy"; sanctions weaken international tourism but promote domestic tourism; sanctions rarely achieve regime change.
Khodadadi (2025)	Brief report / resilience analysis	Iran	Re-imposed U.S. sanctions weakened tourism; resilience strategies include non-Western partnerships, alternative finance, and religious/ domestic tourism.

Authors / year	*Type of paper*	*Case / context*	*Key findings*
Torabi et al. (2024)	Case study	Iran, Shemiranat rural areas	Sanctions and currency devaluation destabilized second-home markets; promoted speculation, corruption, and informal economies.
Ivanov et al. (2017)	Survey (103 hotels)	Russia	Hotels responded with cost-cutting and marketing; impacts varied by size, chain affiliation, and location.
Ovcharov et al. (2015)	Case study	Russia outbound tourism	Sanctions and ruble devaluation triggered outbound tourism crisis, operator bankruptcies, and shifts in demand.
Corbet et al. (2025)	Econometric analysis	Aviation sector (post-Ukraine war)	Sanctions spill over to airlines more than other aviation firms; larger firms hit hardest; uneven regional impacts.
Chung et al. (2024)	fsQCA comparative analysis	China, multiple cases	Outbound tourism sanctions used strategically; effective against tourism-dependent democracies or countries reliant on specific source markets.
Seyfi et al. (2023a)	Interviews / qualitative	Iran	Sanctions constrained mobility, communication, and economic rights; undermined tourism's contribution to SDGs.
Seyfi et al. (2022b)	Interviews / gender lens	Iran	Sanctions eroded women's empowerment in tourism; worsened economic, social, and psychological vulnerabilities.
Seyfi et al. (2025)	Qualitative, feminist lens	Iran	Women entrepreneurs developed "gendered entrepreneurial resilience" as resistance to sanctions and domestic patriarchy.
Toubes & Araújo-Vila (2021)	Content analysis	Travel advisories (Kenya focus)	Language in advisories acted as covert sanctions; persistent use of risk terminology stigmatized destinations.
Li & Zhang (2021)	Case study	North Korea & Dandong (China)	Sanctions disrupted both economies; residents noted tourism decline but also long-term adjustment processes.
Earl (2021)	Methodological reflection	Russia	Sanctions hindered fieldwork access, data collection, and knowledge exchange.
Seyfi et al. (2023b)	Conceptual / geopolitics of health	Global (COVID-19 + sanctions)	Travel bans and sanctions intersected; sanctions worsened pandemic impacts on sanctioned states.
Siyamiyan Gorji et al. (2023)	Interviews	Iran	Sanctions shaped non-visitors' perceptions negatively; visitors reported improved image post-trip; sanctions heightened perceived risks.

Sanctions as multidimensional constraints

Sanctions hold a distinctive position in shaping the connections between politics, economics, and tourism (Seyfi & Hall, 2019). They are political instruments designed to coerce or punish, yet their enforcement works primarily through economic channels such as trade, finance, and mobility restrictions. Tourism, as an industry reliant on cross-border flows of people, capital, and images, becomes especially vulnerable to these measures. The reach of sanctions thus extends beyond diplomatic bargaining into the everyday functioning of destinations, airlines, hotels, and tour operators. In this sense, sanctions not only regulate political relations but also reorder economic exchanges and reshape tourism practices, binding these three spheres together in ways that reveal their mutual dependence.

Sanctions can be broadly divided into four categories, each with different implications for tourism (Seyfi & Hall, 2020a, 2020b, 2020c; Pratt & Alizadeh, 2018) (see also Figure 3.1). *Financial sanctions*, such as restrictions on

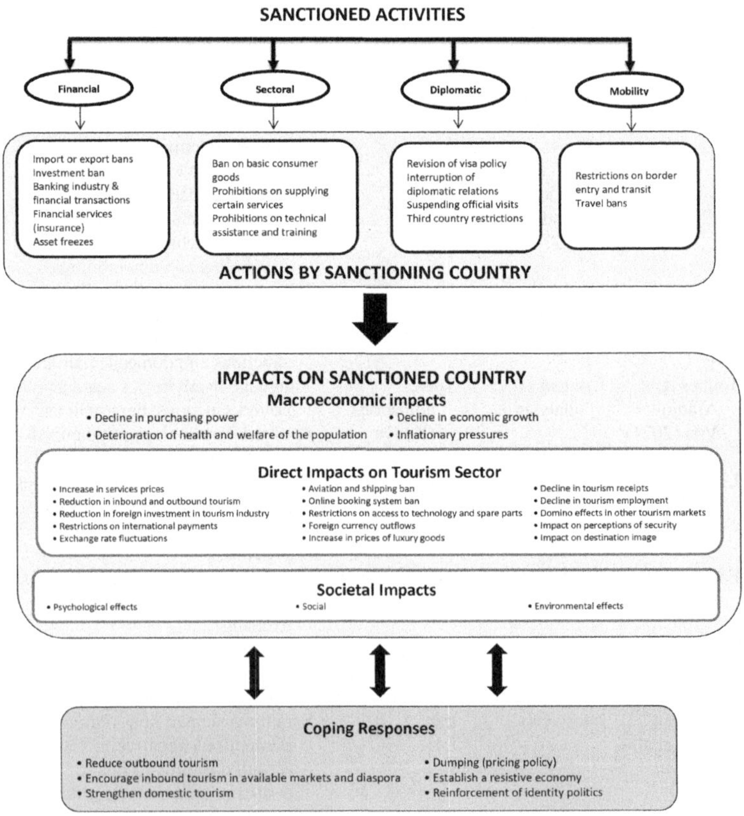

Figure 3.1 Categories of sanctions and their implications for tourism

Source: Seyfi & Hall (2020a)

banking networks or exclusion from systems like SWIFT, undermine international transactions and erode trust in tourism services. *Sectoral sanctions*, targeting industries such as aviation or energy, disrupt critical infrastructures that support travel and hospitality (Box 3.1). *Diplomatic sanctions*, including visa restrictions or travel bans, strike directly at international mobility and reduce flows of visitors. *Individual sanctions*, often directed at elites through asset freezes or travel bans, may seem less directly connected to tourism, but they reinforce the perception of political instability and isolation, thereby affecting tourist decision-making. In addition to these forms, sanctions affect tourism through four types of economic impacts (Pratt & Alizadeh, 2018; Seyfi & Hall, 2020a, 2020b, 2020c). At the macroeconomic level, they contribute to inflation, unemployment, and currency instability, all of which shape the broader conditions in which tourism operates. *Direct* impacts fall on the supply and demand sides of the tourism sector itself, influencing tourist arrivals, hotel occupancy, and airline operations. *Indirect* impacts extend to tourism-supporting industries such as food supply chains, construction, or financial services, which suffer from reduced demand and restricted access to imports. Finally, *induced* impacts reflect the wider distributional effects of sanctions on household income and consumption, reducing local populations' ability to participate in tourism or benefit from its revenues.

Box 3.1 Sanctions and the collapse of Russia's aircraft production

Sanctions imposed on Russia's aviation sector following the 2022 invasion of Ukraine have exposed the fragility of its aircraft manufacturing base. In 2025, Russian firms managed to produce only a single aircraft despite earlier plans to deliver fifteen. Supply chain disruptions, particularly the loss of access to Western parts, technologies, and certification systems, have crippled production lines. A source within the industry admitted that "no component base, no technology, no production facility, or no engineers" remain sufficient to sustain independent operations, and rebuilding these capacities would take "years, if not decades."

The contradiction is striking. While Russian carriers continue to operate large volumes of domestic flights, carrying 111.7 million passengers in 2024, their fleets rely on aging aircraft and indirect import routes to secure spare parts. In many cases, global manufacturers such as Airbus and Boeing fare better than domestic producers, since airlines can still obtain their parts through complex and often informal trade channels.

The consequences extend beyond industry metrics to safety. A series of recent incidents, including the fatal crash of a 1972-built Antonov An-24, highlight the growing risks of operating outdated fleets without reliable access to spare parts or new production. In this sense, sanctions have not only weakened Russia's capacity to compete globally but have also intensified operational hazards in its domestic aviation system.

The Russian case illustrates how sanctions on aviation generate ripple effects across tourism. Aircraft shortages, safety concerns, and reduced connectivity diminish international confidence and limit inbound flows, while outbound markets contract as airfares rise and route options decline. Over time, these pressures risk reinforcing isolation and undermining tourism's potential as an avenue for economic resilience.

Source: Jakobi (2025)

Economic disruptions of sanctions on tourism systems

Sanctions exert some of their most profound impacts on the economic foundations of tourism. Tourism systems are highly internationalized, depending on cross-border financial flows, imported goods and technologies, and sustained foreign investment (Seyfi & Hall, 2019; Siyamiyan Gorji, 2024). When sanctions interrupt these flows, they expose the structural weaknesses of destinations that rely on global integration to maintain competitiveness (Khalid et al., 2024). Iran offers a clear example of this dynamic. After the Islamic Revolution in 1979, the country's tourism sector collapsed, yet between 1995 and 2014, international tourist arrivals still grew at an average annual rate of 14.1%. Scholars argue that this growth could have been much stronger if not for decades of international sanctions, which eroded the national airline fleet, prevented access to aircraft spare parts, and discouraged travelers from key source markets (Jakobi, 2025; Corbet et al., 2025; Khodadadi, 2016). Sanctions also constrained Iran's oil-dependent economy, which traditionally finances around 60% of government revenues and 80% of export earnings, making the wider financial environment unstable and discouraging long-term tourism investment, as firms face higher financing costs, policy uncertainty, and curtailed access to international partners. These conditions delay upgrades to critical infrastructure, depress service quality, and narrow the sector's capacity for sustained growth and diversification.

Investment blockages represent another critical pathway of disruption. International hotel chains, convention centers, and resorts rely heavily on foreign capital, multinational partnerships, and technology transfers. Iran briefly experienced an influx of interest after the 2015 nuclear agreement, with major groups such as Accor entering the market. However, the reimposition of sanctions in 2018 forced abrupt withdrawals, leaving projects incomplete and local partners exposed to significant losses (Khodadadi, 2018). Cuba illustrates the long-term consequences of restricted investment, as U.S. embargoes limited access to credit and construction materials, slowing the modernization of its hospitality sector (Hove et al., 2020). In Russia, Ivanov et al. (2017) found that sanctions eroded hotel revenues, increased input costs, and reduced foreign visitor numbers, with small and medium-sized operators disproportionately affected.

Currency instability further complicates these pressures. Sanctions often trigger devaluations by restricting access to foreign reserves and global financial markets. While weaker currencies can make destinations more affordable for inbound tourists, they simultaneously render outbound travel prohibitively expensive. The collapse of the ruble following the 2022 sanctions sharply curtailed Russian outbound tourism, while the modest gains from inbound flows were insufficient to offset overall losses. Iran's multiple exchange rate systems, shaped by prolonged sanctions, created additional distortions in pricing and investment within the tourism sector (Siyamiyan Gorji et al., 2023). Such volatility not only undermines competitiveness but also encourages speculative behaviors, such as the use of real estate as a hedge against inflation, which destabilize domestic tourism (Torabi et al., 2024).

Financial infrastructures are also among the first to exhibit disruption under sanctions. Modern tourism depends on reliable payment systems, global booking platforms, and electronic transfers. When Iranian banks were excluded from SWIFT in 2012, foreign visitors were unable to use international credit cards, forcing operators to rely on cash transactions (Seyfi & Hall, 2018). This discouraged many tourists, heightened risks of fraud, and destabilized businesses. A comparable situation emerged in Russia after its partial exclusion from SWIFT in 2022, when Visa and Mastercard suspended services. Russian travelers found themselves unable to make payments abroad, while inbound tourists feared being unable to transact within Russia (Ovcharov et al., 2015). Such disruptions undermine confidence and interrupt the basic functionality of tourism operations.

One of the clearest economic consequences of sanctions is the shrinking of tourism markets. Sanctions reduce disposable incomes, widen social inequalities, and restrict the ability of local populations to participate in tourism either as consumers or entrepreneurs. In Iran, for example, sanctions eroded the purchasing power of the middle class, with workers reportedly losing up to 90% of their purchasing capacity over sanction periods (Seyfi & Hall, 2018). These declines reduce domestic demand for tourism services and limit outbound travel, reinforcing isolation. In Russia, the imposition of sanctions following the annexation of Crimea contributed to the collapse of the ruble, weakening both outbound tourism flows and the viability of local operators (Ovcharov et al., 2015). Serbia experienced similar marginalization during the 1990s, when sanctions cut the country off from international markets, stalling long-term development in its tourism sector (Popesku & Hall, 2004).

Gendered impacts of sanctions

The economic pressures generated by sanctions extend beyond aggregate markets and investment flows; they also produce uneven social consequences that are strongly gendered. Sanctions have specific effects on women, intensifying existing inequalities and reshaping their participation in tourism economies. Women are disproportionately concentrated in precarious, low-paid, and informal employment, positions that are often the first to disappear during crises (Jesús Carrasco-Santos et al., 2024; Hosseini et al., 2022; Seyfi et al., 2025). As sanctions reduce incomes, weaken household purchasing power, and destabilize tourism markets, they further limit women's opportunities to engage as workers and entrepreneurs, while at the same time increasing their vulnerability to exclusion and dependency.

Research increasingly shows that sanctions are not gender-neutral. They reinforce pre-existing inequalities, intensify the vulnerabilities faced by women, and reshape their roles within tourism economies. In sanction-affected contexts, women are disproportionately represented in precarious, low-paid, and informal employment, positions that are typically the first to disappear during crises (Seyfi et al., 2022b). One strand of research highlights how sanctions undermine women's empowerment in tourism and hospitality sectors. Seyfi et al. (2022b) show that sanctions in Iran deteriorated women's economic, psychological, social, and political empowerment, with economic empowerment identified as the most fragile foundation. Within a religio-patriarchal system, the loss of financial independence translated into reduced agency in both household and professional domains, exacerbating gender inequality.

A related study by Seyfi et al. (2025) introduces the concept of *gendered entrepreneurial resilience*. Based on women's lived experiences in Iran's tourism sector, it finds that sanctions generate a gender crisis in which women entrepreneurs attempt to resist both external sanctions and restrictive domestic structures. Entrepreneurship becomes a form of independence and survival, yet it is pursued under conditions of heightened precarity. This challenges conventional assumptions of resilience as gender-neutral, showing instead that it is mediated by patriarchal norms and coercive international measures. Perry (2022) broadens the analysis by examining how sanction types matter and indicate that targeted and human rights sanctions, often assumed to be less harmful, can in fact be more damaging to women than comprehensive sanctions. This occurs because targeted sanctions push governments to cut social spending, especially in education, a sector disproportionately relied upon by women. In turn, reduced investment in education undermines both women's opportunities and broader gender equality metrics. Evidence from labor markets supports these observations. Demir and Tabrizy (2022) demonstrate that sanctions disproportionately reduced female employment in Iran's manufacturing industries, especially in sectors dependent on imported inputs or capital-intensive production. The structural fragility of women's employment in such industries mirrors the vulnerabilities seen in tourism and hospitality, where women are concentrated in jobs most exposed to external shocks. At the level of entrepreneurship, Oghazi et al. (2024) adopt a gendered crisis approach to examine Iranian women's nascent entrepreneurial activity. They find that sanctions reduced women's perception of opportunities to start a business more sharply than men's. Although women retained confidence in their skills, sanctions heightened barriers to action, with reduced resources and unstable markets limiting their capacity to translate skills into ventures. Interestingly, the study also shows that sanctions reduced fear of failure among women, suggesting that for some, prolonged exposure to crisis normalizes risk-taking or diminishes expectations of stability. The wider human rights perspective reinforces these patterns. Drury and Peksen (2014) argue that sanctions systematically worsen women's social, political, and economic rights, with the harshest effects observed in poorer states. They contend that sanctions erode institutional capacity and political stability in ways that deepen gender inequality, reducing respect for women's rights across multiple domains. These findings therefore reveal that sanctions reproduce and intensify gender inequalities rather than acting as neutral economic levers. In tourism, this is visible in the loss of employment, declining empowerment, and constrained opportunities for entrepreneurship. What resilience exists is often borne disproportionately by women, who deal with both international restrictions and entrenched domestic hierarchies. The result is a gendered crisis within tourism economies, where women shoulder a double burden: excluded from the benefits of mobility and disproportionately exposed to its restrictions (Carrasco-Santos et al., 2024).

Sanctions and the perceived risk of travel

Sanctions intensify the perception of risk in international tourism, shaping how destinations are evaluated and often discouraging potential visits. Crises in general are known to disrupt travel flows by raising psychological uncertainty, undermining destination attractiveness, and damaging overall image (Akhoondnejad, 2015; Li et al., 2010). Within this context, sanctions are particularly powerful because they portray political instability and insecurity, even when actual safety conditions may not be severe. Tourists, who tend to prefer low-risk destinations, often respond by avoiding sanctioned countries or postponing travel plans (Alvarez & Campo, 2014; Li et al., 2018). Much like the effects of international conflicts on travel risk, sanctions gradually construct an atmosphere of uncertainty. They not only restrict mobility and financial systems but also symbolically communicate instability, elevating perceptions of danger (Khodadadi, 2018; Seyfi & Hall, 2019). Roehl and Fesenmaier (1992) highlighted the importance of destination-specific risks, aligning with Sönmez and Graefe (1998) argument that both actual and perceived risks must be examined in tourism research. However, such perspectives often overlook the heterogeneity of risk perception, as tourists assess risks differently depending on personal characteristics, prior experiences, and destination-specific contexts (Cui et al., 2016).

Recent scholarship has advanced this discussion by showing that sanctions heighten perceived risk across multiple dimensions. Safety and security risks are often the most salient, as sanctions tend to be closely associated with political instability, terrorism, or even the possibility of war (Morakabati, 2011; Siyamiyan Gorji et al., 2023). These concerns directly touch on tourists' sense of personal safety, making them hesitant to visit destinations perceived as conflict-prone. Political risks are equally significant. In sanctioned countries, potential visitors often fear surveillance, restrictions on their freedom of movement, or hostility linked to strained diplomatic relations (Seyfi & Hall, 2019). Such perceptions are magnified when media coverage frames destinations as politically repressive or hostile to outsiders (Khodadadi, 2016).

Concerns about crime risks also emerge in sanctioned contexts. Because sanctions weaken national economies, tourists sometimes assume that higher levels of unemployment and poverty will translate into increased criminal activity, thereby endangering their personal security (Pratt & Alizadeh, 2018). At the same time, financial risks are a central worry for travelers (Morakabati, 2011). The exclusion of sanctioned states from international payment systems such as Visa and Mastercard creates difficulties for booking accommodation, paying for services, or managing money while abroad. Many potential visitors perceive these restrictions as exposing them to the possibility of loss, overpayment, or theft when forced to carry large amounts of cash (Farahani & Shabani, 2014).

In addition, tourists anticipate risks to satisfaction and performance. Satisfaction risks reflect fears that sanctions will diminish the overall quality of the travel experience, whether through limited access to consumer goods, reduced cultural activities, or poor service standards caused by shortages. Closely related are functional or performance risks, which concern the basic reliability of tourism services. Sanctions often disrupt aviation, hospitality, and booking systems, leading potential travelers to expect breakdowns in logistics, difficulties securing transport or accommodation, and restricted access to healthcare and other essential services while abroad (Siyamiyan Gorji et al., 2023).

This multidimensional framework shows that sanctions do not only affect the macro-level reputation of destinations but also the micro-level judgments tourists make when considering travel. Importantly, these risks are perceived differently by visitors and non visitors. Non-visitors frequently report higher levels of insecurity across all dimensions, basing their views largely on media narratives and international political discourse. By contrast, many visitors describe their actual experiences in sanctioned destinations as safer and more welcoming than expected, often noting that conditions were better than the negative images projected abroad. This divergence highlights how risk perception is socially constructed and heavily influenced by information sources rather than direct encounters (Siyamiyan Gorji et al., 2023).

Overall, sanctions transform destinations into risk-laden spaces by embedding multiple layers of uncertainty into the decision-making process. They magnify safety, political, financial, and functional concerns, discouraging mainstream tourists and limiting demand largely to those willing to tolerate higher levels of perceived risk. In this way, sanctions reshape not only the material realities of tourism but also the psychological and symbolic landscapes that determine how destinations are positioned in the global tourism system.

Sanctions, destination image, and country reputation

Sanctions generate reputational consequences that often outweigh their immediate economic effects. Once imposed, they attach powerful symbolic meanings to destinations, portraying them as unstable, unsafe, or politically unacceptable (Khodadadi, 2016; Avraham, 2015). These images are not easily reversed, even if restrictions are later lifted. Tourism depends on positive destination branding that emphasizes hospitality, openness, and safety. Under sanctions, however, this branding is reshaped by narratives of isolation and hostility. Travel advisories play a central role in this process. It is important to recognize that sanctions operate alongside other political and social forces that shape destination images. Political instability, violent conflict, terrorism, or human rights violations can all undermine perceptions of safety and desirability, even in the absence of sanctions (Avraham, 2015). However, sanctions add a distinctive layer of reputational harm by institutionalizing stigma through formal international measures. Unlike episodic events such as protests or localized violence, sanctions construct an enduring label of exclusion that signals to tourists, operators, and investors that a destination lies outside accepted norms of global tourism. This symbolic isolation often proves more persistent than other crises, since even when sanctions are lifted, the negative associations embedded in discourse and advisories remain deeply entrenched. Toubes and Araújo-Vila (2021) show that official advisories frequently employ terms such as "instability," "terrorism," and "threat." These linguistic framings do more than inform travelers. They embed stigma and reinforce negative associations that persist over time.

The case of Iran illustrates how sanctions reshape destination image across cognitive, affective, and conative dimensions. Non-visitors often describe Iran as insecure, unstable, and financially unreliable, perceptions largely shaped by political discourse and media coverage rather than personal experience. Siyamiyan Gorji et al. (2023) showed that from the perspective of most non-visitors, Iran was not only perceived as highly risky but also as less hospitable to Western tourists while sanctions remained in place. Anti-Western sentiment, negative news, bilateral conflicts, terrorism, and personal safety concerns were among the reasons cited for avoidance. This damaged destination image directly undermined behavioral intentions and discouraged visitation. In contrast, visitors who had traveled to Iran often described it in far more positive terms, emphasizing safety, cultural richness, and warm hospitality, and many expressed a willingness to return. This contrast demonstrates how reputational harm is less about lived realities and more about imaginaries produced externally, which are then internalized by non-visitors as reasons for avoidance.

North Korea itself provides another example of how sanctions and security crises interact to damage destination image. As a country heavily sanctioned due to its nuclear testing program, North Korea has long been portrayed as politically isolated, risky, and unwelcoming to outsiders. Li et al. (2018) found that nuclear testing severely damaged Chinese tourists' perceptions of the country, particularly its affective image. These negative effects echoed the consequences of other security-related crises (Sönmez et al., 1998). The study confirmed Beerli and Martín's (2004) argument that political factors and the natural environment strongly influence perceived image. Sanctions intensified these negative associations by institutionalizing North Korea's pariah status in international tourism markets. Yet, interestingly, some Chinese tourists reported that the very conditions of isolation and mystery made the country more attractive, stimulating intentions to visit despite the risks (Buda & Shim, 2015; Li & Ryan, 2015). Even so, in the aftermath of the fourth nuclear test crisis, concerns about regional instability, radiation, and personal safety reduced revisit intentions, showing how sanctions and security crises together can reinforce reputational decline and limit long-term tourism prospects.

Reputational impacts can also extend beyond sanctioned territories, influencing perceptions of neighboring areas. Li and Zhang (2021) found that residents of Dandong, a Chinese border city adjacent to North Korea, associated sanctions not only with the isolation of North Korea but also with a decline in their own city's image. The symbolic exclusion of one country spilled over, discouraging cross-border exchanges and weakening tourism in the surrounding region. These cases are part of a broader pattern observed across many sanctioned destinations.

Summary and conclusion

This chapter has examined how sanctions operate as multidimensional constraints on tourism, shaping outcomes across economic, social, and symbolic domains. The review of existing research showed that sanctions consistently suppress international tourist flows, interrupt investment, and destabilize aviation and hospitality infrastructures. Country-level studies highlighted cycles of disruption and adaptation, as destinations such as Iran, Russia, and Cuba attempted to maintain tourism under shifting restrictions. The chapter also considered how sanctions affect tourism not only as a sector of impact but also as an instrument of statecraft, where outbound bans and travel advisories serve political purposes.

The analysis revealed that sanctions strike most directly at the financial and institutional foundations of tourism systems. They destabilize exchange rates, restrict access to capital and technologies, and undermine trust in financial transactions. These effects, in turn, shrink tourism markets, depress service quality, and weaken the competitiveness of destinations. Importantly, sanctions also produce deeply gendered consequences, disproportionately affecting women workers and entrepreneurs, reducing their empowerment, and forcing forms of resilience that are both precarious and unequal.

Beyond material and gendered consequences, sanctions reshape destination images and reinforce perceptions of risk. They create enduring reputational damage by attaching stigma to sanctioned countries, portraying them as unstable, unsafe, or unwelcoming. These symbolic consequences often outlast the sanctions themselves, demonstrating the long-term implications of geopolitical exclusion for tourism.

Overall, as noted earlier, sanctions are not temporary disturbances but prolonged geopolitical crises that penetrate the very foundations of tourism. Their effects extend from macroeconomic instability to household livelihoods, from airline operations to women's entrepreneurship, and from investment flows to the symbolic standing of destinations in the global tourism system. Recognizing this breadth is vital for developing more critical and inclusive approaches to tourism research, particularly as sanctions remain among the most widely used tools of international politics. The next chapter builds on this discussion by focusing on resilience, examining how destinations, businesses, and communities attempt to adapt to and withstand the pressures of sanctions.

Questions for discussion

- Are the reputational effects of sanctions more damaging to tourism than their direct economic impacts, and if so, why?
- How should we interpret the gap between tourists' perceptions of sanctioned destinations and the lived experiences of those who actually visit?
- Do sanctions unintentionally generate new forms of adaptation and resilience in tourism, or do they primarily entrench vulnerability and exclusion?
- What ethical dilemmas emerge when tourism becomes both a victim of sanctions and an instrument of statecraft?

References

Akhoondnejad, A. (2015). Analyzing the pre-travel, on-travel, and post-travel behaviors of Iran's first-time visitors. *Journal of Travel & Tourism Marketing*, *32*(8), 1023–1033.

Alvarez, M. D., & Campo, S. (2014). The influence of political conflicts on country image and intention to visit: A study of Israel's image. *Tourism Management*, *40*, 70–78.

Avraham, E. (2015). Destination image repair during crisis: Attracting tourism during the Arab Spring uprisings. *Tourism Management*, *47*, 224–232.

Beerli, A., & Martin, J. D. (2004). Factors influencing destination image. *Annals of Tourism Research*, *31*(3), 657–681.

Buda, D. M., & Shim, D. (2015). Desiring the dark: 'A taste for the unusual'in North Korean tourism? *Current Issues in Tourism*, *18*(1), 1–6.

Carrasco-Santos, M. J., Seyfi, S., Hosseini, S., Hall, C. M., Mohajer, B., Almeida-García, F., & Macías, R. C. (2024). Breaking boundaries: Exploring gendered challenges and advancing equality for Iranian women careers in tourism. *Tourism Management*, *103*, 104913.

Chi, P. Y., Yang, Y., & Lin, C. H. (2024). Effect of international travel sanctions on global tourist arrivals. *Scottish Journal of Political Economy*, *71*(5), 649–675.

Chung, J. Y., Paik, W., & Kim, Y. H. (2025). Tourism sanctions as a means of China's economic statecraft: An fsQCA analysis. *Anatolia*, *36*(2), 395–407.

Corbet, S., Nicolau, J. L., & Oxley, L. (2025). Spillover effects of sanctions on airlines. *Journal of Travel Research*, *64*(1), 74–89.

Cui, F., Liu, Y., Chang, Y., Duan, J., & Li, J. (2016). An overview of tourism risk perception. *Natural Hazards*, *82*(1), 643–658.

Demir, F., & Tabrizy, S. S. (2022). Gendered effects of sanctions on manufacturing employment: Evidence from Iran. *Review of Development Economics*, *26*(4), 2040–2069.

Drury, A. C., & Peksen, D. (2014). Women and economic statecraft: The negative impact international economic sanctions visit on women. *European Journal of International Relations*, *20*(2), 463–490.

Earl, A. (2021). Methodological issues in examining sanctions: Reflections on conducting research in Russia. *Tourism Management Perspectives*, *39*, 100858.

Farahani, B. M., & Shabani, M. (2014). The impact of sanctions on Iran's tourism. *International Journal of Resistive Economics*, *2*(1), 1–12.

Hosseini, S., Macias, R. C., & Garcia, F. A. (2022). The exploration of Iranian solo female travellers' experiences. *International Journal of Tourism Research*, *24*(2), 256–269.

Hove, M., Ndawana, E., & Nhemachena, M. A. (2020). How Cuba survived sanctions and the lessons for Zimbabwe. *Jadavpur Journal of International Relations*, *24*(2), 172–195.

Ivanov, S., Sypchenko, L., & Webster, C. (2017). International sanctions and Russia's hotel industry: The impact on business and coping mechanisms of hoteliers. *Tourism Planning & Development*, *14*(3), 430–441.

Jakobi, S. (2025). *How sanctions are halting Russian aircraft production*. https://airlinegeeks.com/2025/08/11/how-sanctions-are-halting-russian-aircraft-production/.

Khalid, U., Okafor, L., & Burzynska, K. (2024). Sanctions and tourist flows: The roles of religion and geography. *Annals of Tourism Research Empirical Insights*, *5*(2), 100143.

Khodadadi, M. (2016). A new dawn? The Iran nuclear deal and the future of the Iranian tourism industry. *Tourism Management Perspectives*, *18*, 6–9.

Khodadadi, M. (2018). Donald Trump, US foreign policy and potential impacts on Iran's tourism industry: Post-nuclear deal. *Tourism Management Perspectives*, *26*, 28–30.

Khodadadi, M. (2025). Tourism resilience in Iran: Navigating sanctions, diplomacy and emerging opportunities. *Research in Hospitality Management*, *15*(2), 102–108.

Li, F., & Ryan, C. (2015). Chinese tourists' motivations and satisfaction of visiting North Korea. *Asia Pacific Journal of Tourism Research*, *20*(12), 1313–1331.

Li, F., Wen, J., & Ying, T. (2018). The influence of crisis on tourists' perceived destination image and revisit intention: An exploratory study of Chinese tourists to North Korea. *Journal of Destination Marketing & Management*, *9*, 104–111.

Li, F. S., & Zhang, G. (2021). Border resident perceptions of sanctions and tourism: A case study of North Korea. *Tourism Management Perspectives*, *38*, 100821.

Li, S., Blake, A., & Cooper, C. (2010). China's tourism in a global financial crisis: A computable general equilibrium approach. *Current Issues in Tourism*, *13*(5), 435–453.

Morakabati, Y. (2011). Deterrents to tourism development in Iran. *International Journal of Tourism Research*, *13*(2), 103–123.

Oghazi, P., Patel, P. C., & Hajighasemi, A. (2024). Gendered crisis approach: Exploring the gendered impact of Iranian sanctions on nascent entrepreneurship outcomes. *International Small Business Journal*, *42*(8), 1016–1046.

Ovcharov, A. O., Ismagilova, G. N., Ziganshin, I. I., & Rysayeva, M. A. (2015). The influence of Western countries' sanctions on the development of the Russian exit tourism. *Asian Social Science*, *11*(11), 289–296.

Perry, K. (2022). Better for whom? Sanction type and the gendered consequences for women. *International Relations*, *36*(2), 151–175.

Popesku, J., & Hall, D. R. (2004). Sustainability as the basis for future tourism development in Serbia. In D. R. Hall (Ed.), *Tourism and transition: Governance, transformation and development* (pp. 95–103). CABI.

Pratt, S. A., & Alizadeh, V. (2018). The economic impact of the lifting of sanctions on tourism in Iran: A computable general equilibrium analysis. *Current Issues in Tourism*, *21*(11), 1221–1238.

Roehl, W. S., & Fesenmaier, D. R. (1992). Risk perceptions and pleasure travel: An exploratory analysis. *Journal of Travel Research*, *30*(4), 17–26.

Seyfi, S., & Hall, C. M. (Eds.). (2018). *Tourism in Iran: Challenges, development and issues*. Routledge.

Seyfi, S., & Hall, C. M. (2019). International sanctions, tourism destinations and resistive economy. *Journal of Policy Research in Tourism, Leisure and Events*, *11*(1), 159–169.

Seyfi, S., & Hall, C. M. (2020a). *Tourism, sanctions and boycotts*. Routledge.

Seyfi, S., & Hall, C. M. (2020b). Sanctions and tourism: Effects, complexities and research. *Tourism Geographies*, *22*(4–5), 749–767.

Seyfi, S., & Hall, C. M. (2020c). Sanctions and tourism: Conceptualisation and implications for destination marketing and management. *Journal of Destination Marketing & Management*, *15*, 100381.

Seyfi, S., Hall, C. M., Saarinen, J., & Vo-Thanh, T. (2023a). Do international sanctions help or inhibit justice and sustainability in tourism? *Journal of Sustainable Tourism*, *31*(12), 2716–2733.

Seyfi, S., Hall, C. M., & Shabani, B. (2023b). COVID-19 and international travel restrictions: The geopolitics of health and tourism. *Tourism Geographies*, *25*(1), 357–373.

Seyfi, S., Hall, C. M., & Vo-Thanh, T. (2022a). Tourism, peace and sustainability in sanctions-ridden destinations. *Journal of Sustainable Tourism*, *30*(2–3), 372–391.

Seyfi, S., Hall, C. M., & Vo-Thanh, T. (2022b). The gendered effects of statecraft on women in tourism: Economic sanctions, women's disempowerment and sustainability? *Journal of Sustainable Tourism*, *30*(7), 1736–1753.

Seyfi, S., Kimbu, A. N., Tavangar, M., Vo-Thanh, T., & Zaman, M. (2025). Surviving crisis: Building tourism entrepreneurial resilience as a woman in a sanctions-ravaged destination. *Tourism Management*, *106,* 105025.

Siyamiyan Gorji, A. S. (2024). *Journey to the country of contrasts: A holistic exploration of the projected and perceived images of Iran as a tourism destination* (Doctoral dissertation, Universidad de Málaga).

Siyamiyan Gorji, A. S., Garcia, F. A., & Mercadé-Melé, P. (2023). Tourists' perceived destination image and behavioral intentions towards a sanctioned destination: Comparing visitors and non-visitors. *Tourism Management Perspectives*, *45*, 101062.

Sönmez, S. F. (1998). Tourism, terrorism, and political instability. *Annals of Tourism Research*, *25*(2), 416–456.

Sönmez, S. F., & Graefe, A. R. (1998). Determining future travel behavior from past travel experience and perceptions of risk and safety. J*ournal of Travel Research*, *37*(2), 171–177.

Torabi, Z. A., Hall, C. M., Allam, Z., Ghaderi, Z., & Sharifi, A. (2024). Sanctions and second home tourism: The experience of rural areas in Shemiranat, Iran. *Journal of Policy Research in Tourism, Leisure and Events*, *16*(4), 620–639.

Toubes, D. R., & Araújo-Vila, N. (2021). The treatment of language in travel advisories as a covert tool of political sanction. *Tourism Management Perspectives*, *40*, 100866.

4 Conceptualizing resilience in sanctions-affected destinations

Introduction

Earlier chapters examined sanctions as instruments of geopolitics and statecraft that reshape mobilities, disrupt economic flows, and transform the image and reputation of destinations. Tourism emerged as one of the most exposed sectors, given its reliance on international connectivity, financial infrastructures, and perceptions of safety and openness. Building on this foundation, the present chapter shifts the focus from disruption to endurance, asking how sanctioned destinations sustain tourism activity when global connections are constrained. The emphasis is not on whether sanctioned destinations recover in the conventional sense but on how tourism systems adapt, persist, and sometimes transform under conditions of prolonged external constraint.

Resilience has become a central concept in tourism crisis studies, applied to contexts ranging from natural disasters and political unrest to health and financial shocks (O'Connor et al., 2025). It is commonly defined as the ability of systems, communities, or businesses to absorb disturbances, reorganize, and continue functioning (Hall et al., 2017; Prayag, 2018). While this definition is useful, its application is not straightforward. Critics argue that it is often treated as an unquestioned good and framed too simplistically, with most studies assuming that crises are short-lived and recovery is primarily a matter of time and effective management (Cretney, 2014).

Sanctions unsettle conventional understandings of resilience by denying the possibility of a stable "pre-crisis" state to which systems might return and by disrupting predictable trajectories of transformation. To engage with these complexities, analysis must extend beyond tourism studies and into wider debates on sanctions. Much of the established literature evaluates credibility and effectiveness through rational-actor models of sender–target interactions (Bapat & Kwon, 2015; Morgan & Kobayashi, 2021). While analytically valuable, such approaches often overlook the differentiated burdens that sanctions impose within societies and their effects on businesses and individuals operating outside elite decision-making (Seyfi et al., 2023a, 2023b). As De Rogatis (2022) argues, resilience in the context of sanctions is not confined to the

DOI: 10.4324/9781003654339-4

adaptive capacity of targeted states. It also depends on the sanctioning coalition's ability to sustain restrictive measures over time without internal fragmentation. For instance, the European Union's policy toward Iran illustrates this point, as member states exhibited varying levels of vulnerability depending on their trade dependencies, demonstrating that resilience under sanctions is relational and contingent, negotiated across states, firms, and communities.

For tourism, resilience cannot be reduced to headline indicators like arrivals or receipts but must be understood as a politically and socially uneven process. Who or what becomes resilient under sanctions? How is resilience mobilized rhetorically by governments, businesses, and communities? And what kinds of futures are enabled or constrained when adaptation is organized under conditions of prolonged duress? These questions guide the discussion in this chapter, which rethinks resilience in sanctioned destinations not as a neutral descriptor of recovery but as a contested practice. The chapter explores how resilience has been conceptualized across disciplines, how it has been applied in tourism studies, and how sanctions expose its limitations. It then examines adaptive strategies at national, community, and business levels, while critically assessing the ethical and political implications of framing endurance as resilience in contexts where survival often conceals deepening precarity.

Crises in tourism: definitions, frameworks, and the place of sanctions

Crises have long been a central theme in tourism research because of the sector's heightened exposure to external shocks and disruptions. They are commonly defined as periods of acute uncertainty that disrupt normal functioning, undermine organizational credibility, and demand urgent responses (Drennan et al., 2014). Yet beyond this functional definition, scholars emphasize their ambiguous and evolving character, noting that crises are rarely linear, often difficult to anticipate, and sometimes indistinguishable from disasters (Glaesser, 2004; Boin et al., 2020). This has led to different conceptualizations: some view crises as discrete events, while others see them as processes that emerge gradually (Roux-Dufort, 2007). To examine these variations, typologies distinguish between fast-burning, cathartic, long-shadow, and slow-burning crises, each with distinct patterns of onset, development, and termination (Boin et al., 2017). The tourism sector has experienced examples of all four, ranging from sudden terror attacks to the gradual erosion of reputations under political instability, which address the need for novel approaches to crisis management.

In response to such disruptions, scholars have developed structured frameworks for crisis management, which typically emphasize planning, response, recovery, reputation repair, and learning (Pournader et al., 2020). For example, Ritchie (2004) proposed three phases of planning, implementation, and evaluation, while Heath (2001) introduced the 4Rs of reduction, readiness, response, and recovery. Coombs (2015) outlined a similar model of pre-crisis,

crisis, and post-crisis stages. These frameworks have been widely applied to contexts such as national crises, terrorism, natural hazards, and pandemics (Becken & Hughey, 2013; Jin et al., 2019). Complementing these models, risk-management approaches emphasize systematic processes of identification, assessment, treatment, and monitoring of threats across financial, operational, strategic, compliance, and reputational domains (Schweizer & Renn, 2019; Norrman & Wieland, 2020; Yang et al., 2021).

While these literatures make important contributions, they face two critical limitations. First, they stress preparedness and role clarity but caution that rigid plans often falter when crises do not align with established templates. For example, prescriptive recovery models designed around natural disasters or terrorist attacks may fail to examine the protracted and externally imposed pressures of sanctions, leaving destinations and firms without suitable guidance. Second, although destination-level approaches provide valuable insight into broad recovery trajectories, they frequently overlook the realities of small firms and micro-level actors. Research shows that smaller tourism businesses, such as family-run guesthouses, local tour operators, and micro-entrepreneurs, face distinct challenges, including limited access to capital, weaker institutional support, and greater vulnerability to external shocks (Mair et al., 2016; Faisal et al., 2020; Liu et al., 2017; Rastegar et al., 2025). For instance, research on tourism micro-enterprises and SMEs in Iran during the COVID-19 pandemic highlighted how limited resources and reliance on informal coping strategies shaped survival in contexts where formal recovery frameworks were inaccessible (Rastegar et al., 2025) (Boxes 4.1 and 4.2). These findings point to the need for resilience frameworks that move beyond sector-wide assumptions to account for firm size, ownership structures, and embeddedness in local economies.

Beyond formal planning models, communication has also emerged as a decisive dimension of crisis management. Scholars highlight how perceptions of threat, uncertainty, and rapid sense-making can shape outcomes (Siyamiyan Gorji et al., 2023). The rise of social media has further transformed this field, increasing speed, reach, and interactivity while simultaneously elevating risks of rumor, misinformation, and loss of message control (Hosseini et al., 2023). Despite these advances, the research record remains incomplete. Much of the tourism crisis literature prioritizes sudden security threats and terrorism, while slower and politically mediated disruptions such as sanctions have received comparatively little attention. Hospitality facilities are particularly exposed, yet sector-specific strategies remain limited. Moreover, recovery is too often measured through aggregate indicators such as visitor volumes or revenue growth. While these metrics examine broad trends, they overlook distributional inequalities across gender, employment status, or firm size (Box 4.3). This narrow focus overlooks who benefits from recovery, who remains vulnerable, and how resilience is unevenly experienced. Sanctions highlight these limitations especially sharply, since their long-term and politicized nature complicates conventional measurements of success.

Box 4.1 Tourism operator mental health and SME resilience during crises

Research on tourism resilience often emphasizes financial preparedness, recovery planning, and adaptation strategies. Yet the lived experiences of small business operators reveal that resilience is not only an organizational characteristic but also deeply intertwined with mental health. A study of tourism SMEs in Victoria, Australia, during the COVID-19 pandemic (Lindsay-Smith et al., 2022) shows how prolonged crises can simultaneously destabilize businesses and deteriorate the psychological wellbeing of owners and managers. Operators reported heightened stress, anxiety, and depression as they faced extended lockdowns, revenue collapse, and continuous uncertainty. The burden was intensified when businesses had to manage staff redundancies, supply shortages, and the fear of infection.

Crucially, the study demonstrates a bidirectional relationship between organizational resilience and mental health. On one hand, resilient businesses, with stronger networks, modest reserves, and collaborative ties, were better able to shield their operators from severe psychological decline. On the other hand, operators with stronger mental health were more capable of making decisions, sustaining engagement, and steering their businesses through prolonged downturns. This dynamic indicates that organizational and individual resilience cannot be separated: both reinforce or undermine each other.

The findings also highlight structural gaps. Access to mental health services for operators, particularly in regional areas, was limited, unevenly distributed, and poorly tailored to small business needs. Many operators relied on self-reliance and informal coping strategies, which offered short-term relief but increased vulnerability over time. Without adequate support, prolonged exposure to stress and financial strain risked burnout and disengagement, threatening not only personal wellbeing but also the long-term viability of local tourism systems.

The lessons from this case resonate with sanction-hit destinations. Just as pandemic restrictions cut off markets and revenues, sanctions generate long-lasting disruptions that place tourism SMEs under intense strain. When resilience is understood narrowly as financial or structural recovery, the psychological dimension is overlooked, despite its centrality to sustaining entrepreneurial commitment and adaptive capacity. A resilience framework attentive to both business continuity and operator wellbeing is therefore essential. Such an approach broadens the discussion of resilience from sector-wide abstractions to the everyday realities of those who carry crises most directly.

Source: After Lindsay-Smith et al. (2022)

Box 4.2 Local business resilience in Çeşme (Turkey) during COVID-19 and the Russia–Ukraine war

Çeşme's long-standing family firms, many operating for more than 25 years, provide a grounded perspective on how locally embedded tourism businesses navigate overlapping shocks. Interviews with 34 owners reveal that resilience in this context is not only a capacity but also a philosophy. Proprietors stressed continuity of craft, community reputation, and staff loyalty as non-negotiables, even during periods of collapsing demand. Business owners framed the pandemic and the war in strikingly different ways. COVID-19 was largely viewed as a finite public-health crisis that could be managed through vaccines, regulations, and gradual reopening. By contrast, the Russia–Ukraine war was perceived as open-ended and geopolitically uncontrollable. This difference shaped emotional responses: the war triggered deeper anxiety about the long-term loss of Russian and Ukrainian visitors, the decline of foreign-exchange revenues, and the risk of a prolonged recovery period.

Despite the challenges, most businesses avoided lay-offs and experimented with low-cost forms of adaptation. Online sales, home delivery, and phone orders for loyal local customers became central coping mechanisms. Product and menu diversification, flexible shifts, discounts, and efforts to build customer trust through hygiene transparency or generous return policies were widely reported. Owners also leveraged their heritage assets, such as craft traditions, culinary practices, or historic buildings, to differentiate themselves online and appeal to authenticity-seeking domestic visitors.

Three enabling factors were especially important for resilience. First, accumulated experiential knowledge from earlier crises fostered prudent cash handling and incremental adjustment. Second, dense local ties with suppliers, repeat customers, and peers created networks of support. Third, targeted government measures, particularly short-term wage and liquidity support, allowed businesses to buy time while reconfiguring their operations. Yet gaps were also visible: limited digital skills and weak technology adoption meant many responses were reactive rather than proactive, exposing a structural vulnerability of micro and small firms. The loss of high-spending Russian and Ukrainian segments forced businesses to rebalance toward domestic and nearby European markets. Owners increasingly turned to social media to highlight distinctiveness and authenticity, although most considered diversification an imperfect substitute for long-haul demand. While such strategies offered breathing space, they could not fully counter structural exposures such as dependence on a few source markets, thin cash reserves, and destination-level risks linked to geopolitics beyond firm control. Owners' determination prevented immediate job losses, but ongoing uncertainty threatens fatigue and under-investment.

Çeşme's experience highlights several lessons for other destinations affected by sanctions or geopolitical instability. Firm-level ingenuity can preserve jobs and capabilities in the short term, but durable resilience depends on complementary public support. Micro-grants, payroll bridges, digital training, and shared e-commerce platforms can strengthen firms' ability to adapt. Destination governance also plays a role by reducing market concentration risks, investing in diversified route development, and promoting heritage-based identities that travel well across digital platforms. In this way, small-scale adaptations inside firms can be connected to broader strategies for resilience at the destination level.

Source: After Nazli (2024)

Box 4.3 Gendered entrepreneurial resilience under sanctions

Recent research on Iranian women tourism entrepreneurs demonstrates that resilience cannot be understood as a gender-neutral process. Instead, it must be seen as deeply contextual, shaped both by the external pressures of international sanctions and by the internal restrictions of a patriarchal and theocratic society. Drawing on social constructionist and poststructural feminist perspectives, Seyfi et al. (2025) introduced the concept of "gendered entrepreneurial resilience" to demonstrate how women entrepreneurs sustain their businesses, assert independence, and resist exclusion under prolonged crisis conditions.

The study reveals that women's resilience strategies extend beyond individual perseverance and involve collective, innovative, and digital practices. Many rely on peer networks and informal associations to share resources, exchange knowledge, and provide emotional support. Collaboration with competitors, sometimes described as "coopetition," helps women stabilize fragile cash flows and collectively cope with resource shortages. Others adopt frugal or bricolage practices, substituting sanctioned imports with local materials and reorienting their products and services toward domestic, diaspora, or regional markets that are less exposed to international restrictions.

Digital tools play an increasingly important role in these resilience practices. Social media, messaging platforms, and online marketplaces allow women to sustain visibility, communicate with clients, and counter negative perceptions of sanctioned destinations. These strategies represent a form of "digital resilience," where entrepreneurship is not only about economic survival but also about reworking industry narratives and asserting professional expertise in highly constrained settings. At the same time, women must carefully navigate restrictive licensing systems, mobility controls, and cultural expectations, often striking a balance between public visibility and compliance with formal rules.

While these mechanisms allow businesses to endure, the study also highlights their limitations. Resilience often takes the form of survival rather than growth, leaving many women trapped in precarious cycles of low margins and burnout. Unequal access to networks or the added responsibilities of caregiving further disadvantage some women. Moreover, policy barriers, including restricted access to finance and rigid licensing frameworks, can undermine even the most resourceful strategies.

Despite these constraints, the study highlights the wider significance of gendered entrepreneurial resilience for tourism systems. By keeping micro-level businesses alive, women contribute to the functioning of tourism economies during periods of geopolitical isolation. Entrepreneurship also acquires symbolic weight, serving as an expression of dignity, independence, and resistance against both sanctions and patriarchal structures.

The findings of their study provide clear implications for policy and practice. Supporting women-led networks and cooperatives, expanding access to microfinance and digital infrastructure, and designing gender-responsive crisis interventions can strengthen resilience in more equitable ways. Partnerships with associations such as local associations of women entrepreneurs offer promising entry points for tailored support. These insights suggest that resilience-building must be understood as both gendered and context-specific, requiring interventions that go beyond technical survival to address entrenched inequalities.

Source: After Seyfi et al. (2025)

Conceptual foundations of resilience

Definitions and scales of resilience across disciplines

Resilience has become a central concept across disciplines, yet its meaning varies depending on the domain and the unit of analysis. A widely cited definition from the UN International Strategy for Disaster Reduction frames it as the capacity of a system, community, or society to withstand, absorb, and adapt to hazards while safeguarding essential structures and functions (UNISDR, 2009). This systems-oriented view has been influential in tourism studies, but scholars increasingly stress that resilience extends beyond short-term recovery to encompass persistence, reorganization, and transformation under stress.

Three influential traditions illustrate these different orientations. *Engineering resilience*, rooted in materials science and stability analysis, conceptualizes resilience as the speed and completeness of return to a prior equilibrium after disturbance (Pimm, 1984). In this view, variability is treated as a problem to minimize, and resilience is measured by efficiency of recovery, for example, how quickly an airport restores operations after a natural disaster (Cimellaro et al., 2010). *Ecological resilience*, by contrast, emphasizes thresholds, diversity, and the possibility of multiple equilibria. Holling (1973) argued that systems can reorganize into different but still viable states, and resilience lies in the ability to absorb shocks without losing fundamental structure and function. Examples include coastal communities adapting to repeated storm surges by redesigning land use and livelihoods rather than attempting to return to pre-disaster conditions. *Evolutionary resilience* moves further by framing resilience as continuous adaptation and innovation in response to changing circumstances. In this perspective, social and economic systems are dynamic, and resilience involves the capacity to chart new trajectories rather than re-establishing old ones (Davoudi et al., 2013). For instance, destinations that use crises to diversify away from mass tourism into smaller-scale or domestic markets reflect evolutionary resilience in practice.

To operationalize these ideas, scholars have employed heuristics such as the adaptive cycle and panarchy. The adaptive cycle illustrates recurring phases of exploitation, conservation, release, and reorganization, showing how systems move through cycles of growth and collapse (Holling, 2001). Panarchy links these cycles across scales, highlighting how rapid, localized dynamics can trigger wider systemic change, as in the way repeated failures of small businesses during a crisis can destabilize regional economies while also opening pathways for new entrants (Walker & Salt, 2012).

Resilience studies have also expanded to multiple levels of analysis. At the *individual* level, psychological research emphasizes traits such as optimism, self-efficacy, and coping strategies that enable people to withstand adversity (Schwarz, 2018). In *organizational* contexts, resilience refers to the ability to anticipate, prepare for, and adapt to both incremental changes and sudden disruptions, often through redundancy, flexibility, and innovation (Prayag et al., 2024, 2020). At the *community* level, resilience reflects collective capacities to respond to and recover from economic downturns or environmental hazards. Examples include regional tourism associations pooling resources to sustain marketing during crises or local networks providing informal support to businesses when formal safety nets are absent. In tourism scholarship, resilience has become an indispensable framework for analyzing how destinations, businesses, and communities respond to repeated crises, from pandemics to environmental disruption and sanctions (O'Connor et al., 2025; Hall et al., 2023; Hall & Naderi Koupaei, 2024; Hall et al., 2025; Hall, 2025; Prayag, 2018; Prayag et al., 2024, 2020; Seyfi et al., 2025). Its rapid uptake reflects a wider turn to adaptation under accelerating global change and also brings familiar challenges of conceptual stretch and normative ambiguity (Amore et al., 2018; Hall, 2025).

Resilience as recovery, transformation, and endurance

Across disciplines, resilience is often framed through metaphors of movement and persistence. Tourism research has drawn heavily on these metaphors, developing three principal interpretations: resilience as *bouncing back* to an earlier state, resilience as *bouncing forward* into transformation, and resilience as *endurance under constraint*. Each framing carries distinct assumptions about time, agency, and possibility, and each offers different insights for understanding how tourism destinations respond to crises. The following subsections examine these three interpretations in more detail, highlighting their differences and implications for tourism under conditions of crisis and sanctions.

Resilience as bouncing back

The earliest and still most widespread framing of resilience is recovery to a pre-shock condition. In ecology, Holling (1973) distinguished resilience from stability, showing how systems could fluctuate yet return to equilibrium. In disaster studies, Manyena (2006) defined resilience as the ability of communities to restore infrastructure, livelihoods, and social functions after disruption. This notion of rapid recovery has been widely adopted in tourism, where resilience is often measured by the restoration of visitor flows, occupancy rates, and revenues (Faulkner et al., 2020). Thailand's recovery after the 2004 Indian Ocean tsunami is frequently cited in these terms. Despite extensive devastation, international arrivals surpassed pre-tsunami levels within three years, supported by coordinated reconstruction, international aid, and aggressive marketing campaigns (Calgaro & Lloyd, 2008). In this context, resilience was equated with rebuilding and returning to "normal." Yet reliance on the bounce-back model is problematic. It assumes that shocks are temporary and that recovery is achievable with sufficient time and investment. As Dogru et al. (2019) show in their quantitative study on climate change, many destinations face structural vulnerabilities where returning to pre-shock conditions is neither possible nor desirable. Similarly, Becken (2013) argues that recovery in the face of climatic disturbances is constrained by thresholds beyond which systems reorganize rather than revert.

Resilience as bouncing forward

A second strand emphasizes transformation. Here, resilience is less about restoring the old order than about reorienting systems toward new trajectories. Holling's (2001) adaptive cycle introduced the idea that crises can trigger phases of release and reorganization, creating room for renewal. Building on this foundation, Folke (2016) framed resilience as the capacity for adaptation and transformation in the face of social and environmental change, while Davoudi et al. (2013) extended this thinking to evolutionary resilience, emphasizing processes of innovation and reorientation in human systems.

In tourism research, these ideas have inspired studies that examine how crises can generate opportunities rather than merely losses. Biggs et al. (2015) showed how destinations recovering from disasters can use disruption to diversify markets, upgrade infrastructure, and embed sustainability principles. Cellini and Cuccia (2019) found that some Italian regions responded to the 2008–2012 financial crisis by shifting toward international markets, demonstrating adaptive rather than purely restorative capacity. Calgaro et al. (2014) synthesized these insights in the Destination Sustainability Framework, which links vulnerability, exposure, and adaptive capacity to highlight how shocks can become catalysts for transformation.

This bouncing forward framing is increasingly prominent in resilience planning debates. Luthe and Wyss (2014) argue that resilience assessment must explicitly integrate adaptive and transformative strategies across governance levels, since incremental recovery is inadequate in the face of accelerating environmental and social change. Chen et al. (2020) also stress the role of agency, showing how communities with greater collective capacity were more successful in reorienting livelihoods after crises. In this context, resilience as bouncing forward shifts attention from recovery to innovation, highlighting crises as potential openings for more sustainable and desirable forms of tourism development. Yet not all crises provide such opportunities, and in the case of sanctions, resilience often takes a different form: one of endurance under constraint.

Resilience as endurance under constraint

Among the many disruptions tourism faces, sanctions represent a distinctive form of crisis, gradually eroding infrastructure, restricting access to key markets, financial systems, and mobility networks, while simultaneously stigmatizing destinations through media narratives and travel advisories (Seyfi & Hall, 2019); in this sense, sanctions constitute a form of "slow violence" against tourism, where disruption is cumulative rather than episodic, ultimately leading to shrinking demand. Resilience in such contexts cannot be conceptualized through conventional framings of "bouncing back" to pre-crisis conditions or "bouncing forward" into transformation. Both framings assume the shock is temporary and that resources are available for recovery or reinvention (Holling, 1973; Hall et al., 2023; Hall, 2025).

The case of sanctions challenges and invites a deeper rethinking of resilience. They unfold gradually, accumulate over time, and remain contingent on international negotiations rather than local recovery efforts (Seyfi & Hall, 2019). This produces what has been described as a "long-lasting crisis," fundamentally different from the temporary shocks usually addressed in tourism research. In such contexts, resilience is better understood as a resistive economy (Seyfi & Hall, 2018), manifesting as patchwork endurance, marked by incremental adjustments and continuous negotiation despite limited resources and shrinking opportunities (Seyfi et al., 2025).

This divergence also reveals the limitations of conventional resilience framings. In contexts such as climate change or financial crises, diversification, innovation, or market flexibility are often cited as measurable signs of adaptation (Dogru et al., 2019; Cellini & Cuccia, 2019). Under sanctions, however, such indicators are far less reliable. Rising arrivals from "friendly" states or modest growth in domestic tourism may register as resilience on paper but often conceal more troubling dynamics, such as dependence on fragile networks or widening inequalities. In practice, many firms and workers rely less on systematic planning than on ad hoc practices, informal economies, and social capital (Rastegar et al., 2025).

At the same time, resilience in sanctioned destinations is highly politicized. Governments often mobilize the discourse of resilience to project sovereignty and legitimacy, framing continued tourism activity as evidence of national strength (De Rogatis, 2022). This stands in contrast to managerial interpretations of resilience in tourism, which emphasize technical fixes such as marketing campaigns, infrastructure investment, or destination governance (Hall et al., 2017; Prayag, 2018). In sanctioned contexts, resilience is less an operational capacity than a political narrative, one that frequently obscures the disproportionate hardships borne by women, informal workers, and small enterprises (Seyfi & Hall, 2019).

Overall, these dynamics suggest that resilience in sanctioned destinations should be theorized not as recovery or transformation but as hybrid adaptation. It blends short-term coping strategies like price discounting, diaspora engagement, and regional targeting with longer-term adjustments to systemic isolation, such as developing parallel payment infrastructures or privileging domestic tourism. Yet these adaptations are often fragile and contingent, producing forms of survival that sustain activity without resolving vulnerability. In this respect, sanctioned destinations reveal the limitations of prevailing resilience frameworks in tourism and point toward the need for approaches that recognize persistence under constraint as a core dimension of resilience.

Market reorientation: reinforcing loyal visitors and cultivating alternatives

A central strategy for destinations under sanctions is to stabilize demand by consolidating ties with established visitor markets while simultaneously cultivating alternatives that are less politically vulnerable. Loyal segments often provide a foundation for resilience. In Iran, pilgrimage tourism to Mashhad and Qom has long ensured a steady flow of Shi'a visitors despite international restrictions, while Cuba relied heavily on Canadian and selected European markets during the U.S. embargo (Seyfi & Hall, 2018). These segments form a buffer against external shocks, offering relatively predictable demand even when mainstream markets contract.

Alongside this consolidation, sanctioned destinations frequently diversify through outreach to "friendly" or geopolitically aligned countries. Russia, China, and regional neighbors often become priority partners, as their governments and travelers face fewer restrictions on engagement. Policy tools play a critical role in enabling this diversification. Visa facilitation, for instance, has been used to encourage flows from target markets: Iran introduced visa-free travel for Chinese tourists and waived visa fees for Iraqis, while Cuba relaxed state-ownership rules to attract investment in tourism infrastructure (Wilson & Látková, 2016) (Box 4.4).

Box 4.4 Cuba's sanction survival

Cuba's half-century under U.S. sanctions offers a clear counterpoint to assumptions that small states inevitably buckle under external pressure. Read through a subaltern realism lens, Havana's strategy combined pragmatic economics and state-led social investment: rapid trade reorientation toward sympathetic partners; a deliberately mixed economy that folded in self-employment and a tightly managed tourism push; hard currency pragmatism (remittances, a dual-currency regime); and aggressive energy-saving plus incremental renewables. These moves were underpinned by disciplined anti-corruption drives and, crucially, priority spending on education and health. The latter underwrote "medical diplomacy," exporting care and know-how in exchange for oil, hard currency, and international goodwill, soft power that repeatedly paid dividends in multilateral arenas. In short, Cuba's path shows how sanctioned states can keep tourism and wider visitor economies viable by pairing market pragmatism with social investment and disciplined governance. For other sanctioned destinations, the practical lesson is not to copy Cuba's model wholesale, but to adapt its core logics: protect public goods, ring-fence tourism revenues for system maintenance, diversify partners and payment systems, and convert service strengths into diplomatic assets.

Source: After Hove et al. (2020)

Regional and diaspora tourism as anchors of resilience

Building on market reorientation strategies, regional and diaspora tourism provide additional stabilizing mechanisms that help sanctioned destinations sustain visitor flows when global markets contract. Proximity and familiarity often make regional tourists less sensitive to geopolitical stigma. Turkey leveraged this by intensifying its marketing to Iranians during periods of Russian sanctions, offering charter flights, rail links, and payment systems that bypassed U.S. restrictions (Seyfi & Hall, 2020a). Similarly, Iran exempted neighboring populations from visa requirements to bolster arrivals from Iraq and Azerbaijan, with religious and medical tourism acting as strong pull factors. Diaspora tourism plays a parallel role, providing relatively stable flows despite negative international publicity (Box 4.5). In Cuba, remittances from diaspora communities have been essential for both household survival and tourism development (Hove et al., 2020), while Iran's large global diaspora represents a market with enduring ties and emotional motivations to return (Cohen & Yefet, 2021).

North Korea offers another example where reliance on regional tourism has cushioned the impact of long-standing sanctions. Chinese tourists, less affected by geopolitical narratives in Western media, account for the overwhelming majority of foreign visitors. Their presence sustains limited tourism flows and helps sustain local employment opportunities, even though international demand remains negligible due to persistent reputational stigma (Li et al., 2018).

Box 4.5 Resilience of Iranian tourism under renewed sanctions

Iran's tourism sector illustrates both the fragility and adaptability of destinations under sustained geopolitical pressure. Following the 2015 JCPOA, the industry expanded rapidly, attracting over 4.8 million international visitors in 2017 and billions in foreign investment. The US withdrawal from the agreement in 2018 reversed these gains, triggering investor flight, currency shortages, and sharp declines in arrivals. The re-election of Donald Trump in 2024 and the restoration of "maximum pressure" sanctions in 2025 further narrowed Iran's international tourism horizons, reinforcing patterns of isolation, reputational risk, and infrastructural stagnation.

Yet resilience strategies have begun to emerge. Iran has actively sought to diversify its markets, strengthening ties with China, Russia, and Persian Gulf states, and promoting religious tourism centred on Mashhad and Qom. Pilgrimage flows from Iraq and Pakistan now form a stabilizing base, providing foreign currency even as Western arrivals collapse. Domestic tourism campaigns, eco-tourism initiatives, and the adaptive reuse of cultural assets such as caravanserais also point to a shift toward more localized and regional tourism economies. At the financial level, the state has experimented with cryptocurrency systems and regional banking networks to circumvent US-controlled payment infrastructure, while some tourism operators have adjusted services to cater to non-Western clientele.

The resilience of Iranian tourism therefore rests not on restoring past connections with Western markets but on cultivating alternative circuits of visitors, finance, and meaning. These strategies carry limits: domestic tourism cannot fully replace international revenues, non-Western markets bring lower per-capita spending, and reputational challenges continue to overshadow Iran's cultural appeal. Still, Iran's case demonstrates that sanctioned destinations retain agency. By leveraging religious and regional flows, experimenting with financial adaptation, and integrating tourism into broader strategies of economic survival, Iran exemplifies how resilience in tourism can be redefined under sanctions. The lesson is not of unqualified success, but of partial continuity, adaptation, and the search for stability in conditions where global access remains uncertain.

Source: After Khodadadi (2025)

Domestic tourism and product diversification

Sanctioned destinations often turn inward, with domestic tourism becoming a critical buffer when international flows decline. Governments typically encourage internal travel by subsidizing infrastructure, promoting national holidays, or restricting outbound travel to retain spending at home. Russia, for example, invested heavily in Crimea and subsidized inland trips after sanctions, while Iran raised departure taxes to discourage outbound tourism (Ovcharov et al., 2015; Seyfi & Hall, 2020b). At the same time, diversification strategies broaden the appeal of destinations. Turkey expanded halal tourism and created new attractions such as underwater archaeological sites, while Iran repositioned itself as a hub for affordable medical tourism, offering specialized services to regional travelers (Hall & Seyfi, 2021). These initiatives show how sanctioned states attempt to stabilize their tourism sectors by cultivating niche segments less sensitive to geopolitical constraints.

Business-level adaptations and pricing strategies

Beyond state-led initiatives, private sector actors develop their own coping strategies. Tour operators and hotels often rely on price discounting, seasonal promotions, or package bundling to retain clientele. In Russia, however, aggressive underpricing and dumping during sanctions led to the bankruptcy of established operators, showing how short-term fixes can undermine long-term resilience by eroding consumer trust (Ovcharov et al., 2015).

For smaller firms, which already operate with limited resources, survival often depends on improvisation, informal networks, and social capital. These strategies help sustain operations under sanctions, yet they remain fragile and highly vulnerable to further shocks (Rastegar et al., 2025).

Managing visibility and bypassing restrictions

Sanctioned destinations also adapt symbolically and logistically to mitigate reputational damage and ease visitor concerns. To shield travelers from secondary penalties, for instance, Iran stopped stamping foreign passports (Hall & Seyfi, 2021). At the same time, governments reframe national narratives through domestic media and promotional campaigns, aiming to counter stigmatization and attract niche audiences. Despite these efforts, credibility and visibility remain deeply constrained. Research shows that such symbolic strategies often operate under severe limitations, making it difficult for sanctioned destinations to shift global perceptions in a meaningful way (Hall & Seyfi, 2021).

Resilience under sanctions: adaptation or legitimized hardship?

While resilience has become a prominent concept in tourism research, its application to sanctioned destinations raises important conceptual and ethical concerns.

At the macro level, resilience is often celebrated when destinations succeed in maintaining visitor flows or stabilizing revenues despite sanctions. Yet such indicators can obscure the social costs. For workers in informal sectors, particularly women and youth, resilience often translates into heightened precarity, wage reductions, job losses, and increased reliance on unpaid labor (Elson, 2010; Seyfi & Hall, 2019). In this sense, resilience discourse risks legitimizing hardship by presenting endurance as success.

Beyond these social costs, a second concern lies in the politicization of resilience. Governments frequently mobilize the rhetoric of resilience to project sovereignty and international defiance (Box 4.6). The ability to attract visitors from "friendly" countries, sustain religious or medical tourism, or grow domestic travel is framed as evidence that sanctions fail. Such narratives overlook structural inequalities, portraying adaptation as voluntary choice rather than constrained necessity. As Seyfi et al. (2025) note, resilience discourse can shift responsibility onto communities while absolving states of deeper structural reform.

A third issue emerges from the limits of existing resilience scholarship. Much of the tourism literature draws on frameworks developed for short-lived crises such as natural disasters or terrorist attacks (Ritchie, 2004; Coombs, 2015). These models emphasize rapid recovery, communication, and market stabilization. Sanctions, however, are protracted, externally imposed, and uncertain in duration. Treating them as analogous to sudden crises underestimates the scale of long-term adjustments, from reconfiguring supply chains to cultivating alternative markets. It also risks romanticizing improvisation and informality as signs of resilience rather than recognizing them as symptoms of systemic exclusion.

Finally, resilience under sanctions is not evenly distributed. Elites and state-linked enterprises may exploit the situation through monopolies and privileged access to scarce resources, while marginalized groups experience intensified vulnerability. Celebrating resilience in such contexts may reinforce authoritarian narratives and concealing widening inequalities.

A more critical approach is therefore required. Rather than simply asking whether destinations are resilient under sanctions, scholarship should interrogate whose resilience is being measured, at what cost, and with what political consequences for the future of tourism.

Box 4.6 Russia's wartime economy: resilience and fragility under sanctions

Since the 2022 invasion of Ukraine, Russia has become the most heavily sanctioned country in the world. Yet its economy has not collapsed. On the contrary, official figures claim growth of 4.3% in 2024, outpacing all G7 economies, driven by unprecedented military spending and redirected energy exports to China and India. The rouble has also performed strongly, bolstered by these flows and by the development of a "shadow fleet" of tankers to evade sanctions.

Beneath these headline figures, however, signs of strain are emerging. Inflation remains close to 10%, interest rates have been lifted to 20%, and the country faces acute labour shortages, with around 2.6 million workers absent due to conscription or emigration. Oil and gas revenues, once the cornerstone of fiscal stability, fell by 35% year-on-year in May 2025, widening the budget deficit and prompting cuts in infrastructure and public services. Analysts warn that the diversion of funds towards the military undermines investment in transport, utilities, and other long-term growth drivers.

Despite record-low unemployment rates (2.3% in 2024), the economy is entering a phase of stagnation. Russian firms struggle to import technology, and sectors such as automotive manufacturing have been badly damaged. While sanctions have not prevented the Kremlin from sustaining the war effort, they constrain prospects for diversification and future development. The EU's commitment to phasing out Russian gas by 2027 further narrows Moscow's options.

Prospects hinge heavily on geopolitics. A peace deal could ease some constraints, particularly if the United States under President Trump normalises trade relations. Yet European sanctions are likely to remain in place, limiting any return to pre-2022 energy ties. Analysts suggest that, although Russia has proved more resilient than expected, the war carries long-term costs that are increasingly difficult to offset.

Source: After Thomas (2025)

Summary and conclusion

This chapter has examined the consequences of sanctions for tourism, highlighting how measures designed as foreign policy tools reverberate across economies, industries, and communities. Building on earlier chapters that explored sanctions' economic and gendered impacts, the discussion here has shown how tourism systems are especially sensitive to disruptions in finance, mobility, and perception. Financial sanctions restrict international transactions, sectoral sanctions disrupt aviation and energy, and diplomatic sanctions reduce flows of visitors. Even when sanctions target elites, they contribute to broader narratives of instability, discouraging travel and investment.

At the same time, the analysis has revealed that sanctions are not evenly experienced. Destination-level frameworks often chart recovery trajectories but overlook the realities of small firms and micro-enterprises, which face specific vulnerabilities due to limited capital reserves, reliance on informal coping mechanisms, and weaker institutional support. Women are disproportionately represented in such positions, meaning that sanctions reproduce and deepen pre-existing inequalities.

The resilience literature offers useful but limited tools for interpreting these processes. Engineering, ecological, and evolutionary strands shed light on how systems respond to shocks, but they often overlook the social and political dimensions of sanctions. For instance, the adaptive cycle heuristic illustrates how crises can spark reorganization, yet it does not reflect the fact that some groups, especially women, lack equal access to resources that would enable adaptation. Likewise, while tourism resilience frameworks stress preparedness, role clarity, and institutional learning, they often remain silent on how geopolitical decisions and international coalitions shape the scope of possible recovery.

Overall, the evidence presented suggests that sanctions not only weaken tourism economies but also redefine their trajectories. They close off certain pathways, while opening others, such as domestic or regional tourism, digital adaptation, and collective empowerment strategies. Yet these shifts may entrench unevenness, benefiting some actors while marginalizing others. Claims of resilience must therefore be approached critically: survival strategies may conceal deeper vulnerabilities, and what appears adaptive in the short term may reinforce dependence or exclusion over time.

Questions for discussion

1. To what extent can tourism resilience frameworks adequately reflect the specific pressures generated by sanctions, as opposed to other types of crises?
2. Do alternative markets and financial systems provide genuine pathways for resilience, or do they risk deepening dependency on politically unstable or asymmetrical partnerships?
3. How should researchers and policymakers assess the gendered dimensions of resilience, particularly when women entrepreneurs rely on informal coping strategies that may sustain businesses but also perpetuate precarity?
4. What lessons can be drawn from sanctioned destinations such as Iran, Cuba, or Russia for tourism systems elsewhere that may face future geopolitical shocks?
5. Can the notion of "gender-transformative resilience" move beyond survival and adaptation to challenge the structural conditions that sanctions exacerbate?

References

Amore, A., Prayag, G., & Hall, C. M. (2018). Conceptualizing destination resilience from a multilevel perspective. *Tourism Review International, 22*(3–4), 235–250.

Bapat, N. A., & Kwon, B. R. (2015). When are sanctions effective? A bargaining and enforcement framework. *International Organization, 69*(1), 131–162.

Becken, S. (2013). Developing a framework for assessing resilience of tourism sub-systems to climatic factors. *Annals of Tourism Research, 43*, 506–528.

Becken, S., & Hughey, K. F. (2013). Linking tourism into emergency management structures to enhance disaster risk reduction. *Tourism Management, 36*, 77–85.

Biggs, D., Hicks, C. C., Cinner, J. E., & Hall, C. M. (2015). Marine tourism in the face of global change: The resilience of enterprises to crises in Thailand and Australia. *Ocean & Coastal Management, 105*, 65–74.

Boin, A., Ekengren, M., & Rhinard, M. (2020). Hiding in plain sight: Conceptualizing the creeping crisis. *Risk, Hazards & Crisis in Public Policy, 11*(2), 116–138.

Boin, A., Hart, P. T., Stern, E., Stern, E., & Sundelius, B. (2017). *The politics of crisis management*. Cambridge University Press.

Calgaro, E., & Lloyd, K. (2008). Sun, sea, sand and tsunami: Examining disaster vulnerability in the tourism community of Khao Lak, Thailand. *Singapore Journal of Tropical Geography*, *29*(3), 288–306.

Calgaro, E., Lloyd, K., & Dominey-Howes, D. (2014). From vulnerability to transformation: A framework for assessing the vulnerability and resilience of tourism destinations. *Journal of Sustainable Tourism*, *22*(3), 341–360.

Cellini, R., & Cuccia, T. (2019). Do behaviours in cultural markets affect economic resilience? An analysis of Italian regions. *European Planning Studies*, *27*(4), 784–801.

Chen, F., Xu, H., & Lew, A. A. (2020). Livelihood resilience in tourism communities: The role of human agency. *Journal of Sustainable Tourism*, *28*(4), 606–624.

Cimellaro, G. P., Reinhorn, A. M., & Bruneau, M. (2010). Framework for analytical quantification of disaster resilience. *Engineering Structures*, *32*(11), 3639–3649.

Cohen, R. A., & Yefet, B. (2021). The Iranian diaspora and the homeland: Redefining the role of a centre. *Journal of Ethnic and Migration Studies*, *47*(3), 686–702.

Coombs, W. T. (2015). The value of communication during a crisis: Insights from strategic communication research. *Business Horizons*, *58*(2), 141–148.

Cretney, R. (2014). Resilience for whom? Emerging critical geographies of socio-ecological resilience. *Geography Compass*, *8*(9), 627–640.

Davoudi, S., Brooks, E., & Mehmood, A. (2013). Evolutionary resilience and strategies for climate adaptation. *Planning Practice & Research*, *28*(3), 307–322.

De Rogatis, P. (2022). Resilience in economic sanctions: The neglected relationship between resiliency and credibility in the EU. *Sentio*, *4*, 12–25.

Dogru, T., Marchio, E. A., Bulut, U., & Suess, C. (2019). Climate change: Vulnerability and resilience of tourism and the entire economy. *Tourism Management*, *72*, 292–305.

Drennan, L. T., McConnell, A., Stark, A., & Drennan, L. T. (2014). *Risk and crisis management in the public sector*. Routledge.

Elson, D. (2010). Gender and the global economic crisis in developing countries: A framework for analysis. *Gender & Development*, *18*(2), 201–212.

Faisal, A., Albrecht, J. N., & Coetzee, W. J. (2020). Renegotiating organisational crisis management in urban tourism: Strategic imperatives of niche construction. *International Journal of Tourism Cities*, *6*(4), 885–905.

Faulkner, J. P., Murphy, E., & Scott, M. (2020). Developing a holistic 'vulnerability-resilience' model for local and regional development. *European Planning Studies*, *28*(12), 2330–2347.

Folke, C. (2016). Resilience (republished). *Ecology and Society*, *21*(4).

Glaesser, D. (2004). *Crisis management in the tourism industry*. Routledge.

Hall, C. M. (2025). On flickering, tipping points, trajectories, transitions, transformations, regime shifts, and resiliences in tourism: Why resilient does not mean better and why change is not always sustainable. *Via. Tourism Review*, (27).

Hall, C. M., & Naderi Koupaei, S. (2024). The possibilities of resilience for service organisations. *Journal of Services Marketing*, *38*(4), 397–403.

Hall, C. M., Prayag, G., & Amore, A. (2017). *Tourism and resilience: Individual, organisational and destination perspectives*. Channel View Publications.

Hall, C. M., Saarinen, J., & Seyfi, S. (2025). Rethinking resilient futures for Nordic tourism. *Scandinavian Journal of Hospitality and Tourism*, *24*(4–5), 375–389.

Hall, C. M., Safonov, A., & Naderi Koupaei, S. (2023). Resilience in hospitality and tourism: Issues, synthesis and agenda. *International Journal of Contemporary Hospitality Management*, *35*(1), 347–368.

Hall, C. M., & Seyfi, S. (2021). Tourism and sanctions. In *Research handbook on economic sanctions* (pp. 351–368). Edward Elgar Publishing.

Heath, R. (2001). Effective problem and crisis management-FAST© and the six interaction rules. *International Journal of Emergency Management*, *1*(1), 49–60.

Holling, C. S. (1973). Resilience and stability of ecological systems. *Annual Review of Ecology and Systematics*, *4*(1), 1–23.

Holling, C. S. (2001). Understanding the complexity of economic, ecological, and social systems. *Ecosystems*, *4*(5), 390–405.

Hosseini, S., Siyamiyan Gorji, A., Vo-Thanh, T., & Zaman, M. (2023). Behind the hashtags: Exploring tourists' motivations in a social media-centred boycott campaign. *Tourism Recreation Research*. https://doi.org/10.1080/02508281.2023.2210010.

Hove, M., Ndawana, E., & Nhemachena, M. A. (2020). How Cuba survived sanctions and the lessons for Zimbabwe. *Jadavpur Journal of International Relations*, *24*(2), 172–195.

Jin, X. C., Qu, M., & Bao, J. (2019). Impact of crisis events on Chinese outbound tourist flow: A framework for post-events growth. *Tourism Management*, *74*, 334–344.

Khodadadi, M. (2025). Tourism resilience in Iran: navigating sanctions, diplomacy and emerging opportunities. *Research in Hospitality Management*, *15*(2), 102–108.

Li, F., Wen, J., & Ying, T. (2018). The influence of crisis on tourists' perceived destination image and revisit intention: An exploratory study of Chinese tourists to North Korea. *Journal of Destination Marketing & Management*, *9*, 104–111.

Lindsay-Smith, G., Pyke, J., Gamage, A., Nguyen, V., & De Lacy, T. (2022). Tourism operator mental health and its relationship with SME organisational resilience during disasters. *Tourism Management Perspectives*, *42*, 100961.

Liu, Y., Shankar, V., & Yun, W. (2017). Crisis management strategies and the long-term effects of product recalls on firm value. *Journal of Marketing*, *81*(5), 30–48.

Luthe, T., & Wyss, R. (2014). Assessing and planning resilience in tourism. *Tourism Management*, *44*, 161–163.

Mair, J., Ritchie, B. W., & Walters, G. (2016). Towards a research agenda for post-disaster and post-crisis recovery strategies for tourist destinations: A narrative review. *Current Issues In Tourism*, *19*(1), 1–26.

Manyena, S. B. (2006). The concept of resilience revisited. *Disasters*, *30*(4), 434–450.

Morgan, T. C., & Kobayashi, Y. (2021). Talking to the hand: Bargaining, strategic interaction, and economic sanctions. *European Economic Review*, *134*, 103685.

Nazli, M. (2024). Adaptation of local businesses to the new era during the COVID-19 and the Russia-Ukraine war: Case of Çeşme. *Current Issues in Tourism*, *27*(8), 1173–1184.

Norrman, A., & Wieland, A. (2020). The development of supply chain risk management over time: Revisiting Ericsson. *International Journal of Physical Distribution & Logistics Management*, *50*(6), 641–666.

O'Connor, P., Esfandiar, K., & Hallak, R. (2025). Resilience for hospitality and tourism enterprises: A review and directions for future research. *Tourism Analysis*, *30*(1), 5–22.

Ovcharov, A. O., Ismagilova, G. N., Ziganshin, I. I., & Rysayeva, M. A. (2015). The influence of Western countries' sanctions on the development of the Russian exit tourism. *Asian Social Science*, *11*(11), 289–296. doi:10.5539/ass.v11n11p289

Pimm, S. L. (1984). The complexity and stability of ecosystems. *Nature*, *307*(5949), 321–326.

Pournader, M., Kach, A., & Talluri, S. (2020). A review of the existing and emerging topics in the supply chain risk management literature. *Decision Sciences*, *51*(4), 867–919.

Prayag, G. (2018). Symbiotic relationship or not? Understanding resilience and crisis management in tourism. *Tourism Management Perspectives*, *25*, 133–135.

Prayag, G., Ozanne, L. K., & Chowdhury, M. (2024). How can service organizations build resilience by leveraging capabilities and service worker team knowledge? *Journal of Services Marketing*, *38*(4), 474–493.

Prayag, G., Spector, S., Orchiston, C., & Chowdhury, M. (2020). Psychological resilience, organizational resilience and life satisfaction in tourism firms: Insights from the Canterbury earthquakes. *Current Issues in Tourism*, *23*(10), 1216–1233.

Rastegar, R., Seyfi, S., & Shahi, T. (2025). Tourism SMEs' resilience strategies amidst the COVID-19 crisis: The story of survival. *Tourism Recreation Research*, *50*(2), 428–434.

Ritchie, B. W. (2004). Chaos, crises and disasters: A strategic approach to crisis management in the tourism industry. *Tourism Management*, *25*(6), 669–683.

Roux-Dufort, C. (2007). Is crisis management (only) a management of exceptions? *Journal of Contingencies and Crisis Management*, *15*(2), 105–114.

Schwarz, S. (2018). Resilience in psychology: A critical analysis of the concept. *Theory & Psychology*, *28*(4), 528–541.

Schweizer, P. J., & Renn, O. (2019). Governance of systemic risks for disaster prevention and mitigation. *Disaster Prevention and Management: An International Journal*, *28*(6), 862–874.

Seyfi, S., & Hall, C. M. (Eds.). (2018). *Tourism in Iran: Challenges, development and issues*. Routledge.

Seyfi, S., & Hall, C. M. (2019). International sanctions, tourism destinations and resistive economy. *Journal of Policy Research in Tourism, Leisure and Events*, *11*(1), 159–169.

Seyfi, S., & Hall, C. M. (2020a). *Tourism, sanctions and boycotts*. Routledge.

Seyfi, S., & Hall, C. M. (2020b). Sanctions and tourism: Conceptualisation and implications for destination marketing and management. *Journal of Destination Marketing & Management*, *15*, 100381.

Seyfi, S., Hall, C. M., Saarinen, J., & Vo-Thanh, T. (2023a). Do international sanctions help or inhibit justice and sustainability in tourism? *Journal of Sustainable Tourism*, *31*(12), 2716–2733.

Seyfi, S., Hall, C. M., & Shabani, B. (2023b). COVID-19 and international travel restrictions: The geopolitics of health and tourism. *Tourism Geographies*, *25*(1), 357–373.

Seyfi, S., Kimbu, A. N., Tavangar, M., Vo-Thanh, T., & Zaman, M. (2025). Surviving crisis: Building tourism entrepreneurial resilience as a woman in a sanctions-ravaged destination. *Tourism Management*, *106*, 105025.

Siyamiyan Gorji, A. S., Garcia, F. A., & Mercadé-Melé, P. (2023). Tourists' perceived destination image and behavioral intentions towards a sanctioned destination: Comparing visitors and non-visitors. *Tourism Management Perspectives*, *45*, 101062.

Thomas, D. (2025). Russia's economy is down but not out. *BBC*. https://www.bbc.com/news/articles/cpqnj17vryjo.

UNISDR. (2009). Terminology on disaster risk reduction. *United Nations*. http://unisdr.org/files/7817_UNISDRTerminologyEnglish.pdf.

Walker, B., & Salt, D. (2012). *Resilience thinking: Sustaining ecosystems and people in a changing world*. Island press.

Wilson, J., & Látková, P. (2016). American tourists in Cuba: Is the communist country a forbidden or a tainted fruit? *Tourism and Hospitality Research*, *16*(1), 6–18.

Yang, J., Xie, H., Yu, G., & Liu, M. (2021). Antecedents and consequences of supply chain risk management capabilities: An investigation in the post-coronavirus crisis. *International Journal of Production Research*, *59*(5), 1573–1585.

5 Toward resilient tourism futures in sanctions-ridden destinations

Introduction

The previous chapters have shown that sanctions are no longer marginal or exceptional; they have become one of the defining instruments of twenty-first-century geopolitics, shaping the flows, economies, and images that underpin global tourism. Where earlier generations of tourism scholarship largely concentrated on natural hazards, pandemics, terrorism, or economic downturns, the analysis here has demonstrated that sanctions deserve recognition as a distinct form of crisis. Unlike sudden disruptions with identifiable beginnings and ends, sanctions are deliberately imposed, externally driven, and often indefinite. They introduce constraints that are cumulative, deeply politicized, and outside the control of destinations themselves. For tourism, this shift is profound. Sanctions reach into the most basic infrastructures of mobility and exchange. They determine who can cross borders, which airlines can operate, whether payments can be processed, and how destinations are imagined in the global marketplace. A single decision taken in Washington, Brussels, or New York can ripple through hotels in Havana, tour operators in Tehran, or airlines in Yangon. These are not incidental shocks but systemic blockages that reshape the very conditions under which tourism is possible.

The case studies of Iran, Cuba, Russia, Turkey, and Myanmar have highlighted the uneven and often contradictory ways in which tourism actors cope with sanctions. In some contexts, diaspora networks or alternative markets provide partial lifelines. In others, domestic tourism becomes a buffer, even if it cannot replace lost international revenues. Entrepreneurs improvise with cash-based economies or informal supply chains, while workers bear the brunt of precarity through lower wages, lost jobs, or unpaid labor. Governments, meanwhile, deploy the language of resilience as a political instrument, celebrating endurance as proof of sovereignty, yet this narrative often overlooks the hardships borne by communities and individuals.

Viewed in this context, the pattern of sanctions points to a pressing need to rethink what resilience means in a sanctioned world. If resilience is reduced to arrivals data or the survival of firms, then sanctioned destinations may appear

DOI: 10.4324/9781003654339-5

superficially stable. But such stability is often sustained by deepening inequality, exploitative arrangements, and exclusion from global circuits. To evaluate resilience under sanctions requires a critical lens that asks: resilient for whom, and at what cost? What futures are made possible by adaptation, and which are obstructed by prolonged isolation?

This chapter takes up these questions by drawing together the insights of the preceding analysis and considering their wider implications for both scholarship and policy. It interrogates the ways resilience has been framed, highlighting its tendency to obscure uneven impacts and reproduce official narratives of endurance, while overlooking the lived experiences of tourism workers, entrepreneurs, and communities. It also identifies the limitations of existing research, which has often examined sanctions as background conditions rather than as central forces shaping tourism systems. By critically reassessing the conceptual and empirical foundations of resilience in sanctioned destinations, the chapter aims to map future research directions and to outline what a more inclusive and politically aware approach to resilience might look like.

Rethinking resilience for sanctioned destinations

The preceding case studies demonstrate that the dominant framing of resilience in tourism, as the ability to "bounce back" after a discrete shock, is ill-suited to contexts shaped by sanctions. Sanctions do not end with reconstruction or vaccine rollout; they persist as structural constraints that reshape economies and societies over years, even decades (Cortright & Lopez, 2018; Seyfi & Hall, 2019). For sanctions-ridden destinations, there is no baseline to return to. Instead, resilience is better understood as persistence under constraint, a continual negotiation with scarcity, uncertainty, and externally imposed barriers (De Rogatis, 2022; Khodadadi, 2025).

Improvised and uneven coping strategies are central to this persistence. Households often rely on remittances from diasporas, businesses experiment with barter or cryptocurrency, and governments substitute domestic markets for lost international demand. These measures sustain survival but rarely enable recovery (Hall & Seyfi, 2021). They are fragmented, localized, and shaped by uneven access to resources. In Iran, for instance, tour operators with diaspora connections could secure modest flows of visitors even after sanctions tightened, while small family-run guesthouses without such networks struggled to survive (Oghazi et al., 2024; Seyfi et al., 2025). Similarly, Cuba's reliance on Canadian and European tourists demonstrates adaptive diversification (Hove et al., 2020), but it also highlights dependence on a few key markets, leaving the system highly vulnerable to external shocks.

A critical concern is the co-option of resilience narratives by states. Governments in sanctioned destinations often appropriate the language of resilience to project self-sufficiency and legitimacy. In Russia, campaigns promoting "patriotic tourism" framed domestic travel as a patriotic duty, even as local businesses faced declining revenues and shrinking consumer bases (Kozlovskaya & Kozlovskaya, 2021). Such narratives can overlook deepening inequalities and silence dissent, recasting endurance under hardship as evidence of national strength. In these cases, resilience becomes less a lived condition of communities and more a political tool of states, raising questions about whose resilience is being celebrated and whose suffering is being overlooked. Justice and equity therefore become central evaluative lenses in studying resilience under sanctions. The uneven distribution of burdens is stark: women, informal workers, and peripheral regions often absorb the greatest costs, while elites with access to networks and capital may benefit from monopolized opportunities created by scarcity (Seyfi et al.,2023a). Rarick (2006) raises a stark ethical dilemma, whether it is defensible to dismantle a society in the name of saving it. The resolution of such dilemmas, though often disputed, tends to be shaped at both national and global levels. These ethical questions guide the following discussion of sanctions. If resilience is measured only through conventional indicators such as arrivals, occupancy rates, or revenues, such measures may conceal deep inequalities. A more critical approach presses further: Who gains from strategies of adaptation? Who is marginalized? What compromises are imposed to maintain continuity in tourism? Reframing resilience in this way shifts attention from abstract system survival to the social, political, and ethical dimensions of endurance under sanctions. It opens the way for a more honest discussion about tourism futures in sanctioned destinations, not only how systems endure, but at what cost, and with what consequences for justice and inclusion.

Research concerns on resilience in an era of sanctions

Although resilience has become a central concept in tourism studies, its application to sanctioned destinations exposes several weaknesses in the existing literature. One concern is conceptual slippage. Too often, resilience is treated as a catch-all term that refers indiscriminately to any form of coping or persistence. As Espinoza-Figueroa et al. (2025) showed in the case of Ecuador during the COVID-19 crisis, many strategies celebrated as resilience were in fact tactical survival measures that left underlying structural vulnerabilities intact. In sanctioned contexts, this danger is even sharper. A destination may appear resilient in the sense that visitor arrivals stabilize or firms continue to operate (Khodadadi, 2025), yet such stability may conceal widening inequality, exploitative labor conditions, and forced informalization (Seyfi et al., 2025). Moreover, resilience narratives are frequently mobilized by states as political instruments. Governments often promote endurance under sanctions as evidence of sovereignty or defiance, but this discursive use glosses over the uneven burdens borne by workers, small-scale entrepreneurs, and marginalized communities. Distinguishing between state-promoted resilience and lived resilience remains a significant research challenge.

Another concern lies in the overreliance on formal indicators. Much tourism resilience research evaluates performance through quantifiable metrics such as visitor numbers, hotel occupancy, or revenue recovery (e.g., Zhang et al., 2024). In sanctioned economies, however, many adaptive strategies emerge in informal and opaque spaces: cash-based transactions, barter systems, or parallel supply chains. These practices are essential to survival but are rarely examined in official statistics. An added difficulty is that sanctioned governments often control, distort, or suppress tourism data for political ends. Reliance on formal indicators can overlook the hidden but vital practices that sustain livelihoods. Addressing this gap calls for more ethnographic, qualitative, and community-centered approaches that examine everyday practices of coping and improvisation.

A third issue concerns the temporal and scalar framing of resilience. Most resilience studies adopt short- to mid-term horizons, focusing on recovery over several months or years (Rastegar et al., 2025). Yet sanctions often endure for decades, producing slow-moving processes of decline that fall outside these analytical windows. Long-term effects such as capital erosion, reputational damage, and brain drain reshape destinations far beyond the scope of typical resilience studies (Siyamiyan Gorji et al., 2023). Scale is also problematic. Analyses often concentrate on firms or destinations while neglecting the ways sanctions are shaped by global geopolitics. The decision of a single actor in Washington or Brussels can restructure entire industries in Tehran or Havana, meaning that resilience at the micro level cannot be understood apart from these macro-level forces.

Resilience research also tends to treat actors as if they operate on equal footing, thereby under-theorizing inequality and differential vulnerability. Yet sanctioned contexts are marked by stark disparities in adaptive capacity. Women, rural communities, informal workers, and minority groups are disproportionately exposed to precarity, often without access to credit, institutional protection, or mobility (Oghazi et al., 2024; Demir & Tabrizy, 2022). Without explicit attention to these power asymmetries, resilience can become an abstract ideal that conceals the deepening of inequality.

There is also a tendency for resilience scholarship to lapse into policy instrumentalism by offering lists of best practices that may be relevant to sudden shocks but prove shallow under sanctions. Prescriptions such as diversify markets, promote domestic tourism, or adopt digital platforms, assume conditions that are often absent in sanctioned economies (Khodadadi, 2025). Domestic tourism is not a reliable buffer when inflation and currency collapse erode disposable income. Calls for digital innovation are hollow when global payment systems are inaccessible and websites are blocked. Such prescriptions may even exacerbate inequality by favoring urban elites or tech-savvy firms while leaving smaller actors behind. More critical policy analysis is needed to address structural barriers, institutional weaknesses, and geopolitical constraints.

Finally, sanctions highlight the neglect of transitional and exit pathways in resilience research. Much of the literature treats resilience as indefinite persistence or cyclical recovery. Yet in sanctioned economies, the question of transition is central. How can destinations re-enter global markets? What strategies enable reputational repair, reintegration, or reconfiguration of alliances? Very few tourism studies consider these dimensions, despite their importance for long-term recovery. By focusing narrowly on adaptation, research can end up naturalizing sanctions as a permanent condition rather than interrogating pathways toward transformation.

Concluding reflections: resilience, sanctions, and unanswered questions

This book was motivated by a simple but pressing puzzle: why has tourism research paid so little attention to sanctions when these measures have become one of the defining instruments of twenty-first-century geopolitics? While destinations shaken by earthquakes, pandemics, or terrorism have received extensive scholarly attention, sanctioned societies, often cut off from global mobility and financial systems for years at a time, have remained at the margins of tourism studies. The impulse to write this book emerged from conversations with entrepreneurs in sanction-ridden destinations, many of whom posed a similar concern: why does the world discuss our survival only when bombs fall or diseases spread, but not when politics quietly strangles the economy year after year? It is this silence that this book set out to challenge.

To address this gap, the book engaged three interrelated questions. First, how do sanctions operate as a distinct form of tourism crisis, different from sudden shocks, yet no less destructive? Second, how do destinations, businesses, and workers adapt to externally imposed isolation, sometimes creatively but often precariously? Third, what does resilience mean when the very infrastructures of mobility, airlines, payments, visas, and images, are systematically constrained by geopolitical power? These questions guided the analysis across the chapters, which combined case studies of Iran, Cuba, Russia, Turkey, and Myanmar with broader conceptual debates on resilience, adaptation, and political economy.

Each chapter contributed a part of the wider argument. The early chapters mapped the theoretical terrain, showing how conventional definitions of resilience, often borrowed from engineering or ecology, struggle to capture the deliberate, indefinite, and politicized nature of sanctions. The case studies illustrated how adaptation unfolds unevenly: through diaspora networks in Cuba, cash-based improvisations in Iran, religious corridors between Iraq and Iran, or digital circumvention strategies in Russia. Later chapters examined how the very language of resilience can be appropriated by governments, not as a neutral concept, but as a political instrument that celebrates endurance while concealing the burdens shifted onto vulnerable groups. The final chapters synthesized these insights, arguing that resilience under sanctions must be rethought not as a technical attribute of systems but as a contested and deeply political process.

The book does not aim to offer neat solutions. Instead, it seeks to unsettle prevailing assumptions. If resilience is measured only by arrival statistics or GDP contributions, sanctioned destinations may appear superficially stable. Yet, as this analysis has shown, such stability is often sustained by deepening inequality, heightened precarity, and prolonged exclusion from global circuits. By placing sanctions at the center of tourism scholarship, the book argues that they should not be treated as exceptional disruptions but as systemic conditions that fundamentally reshape how tourism is imagined and practiced.

At the same time, many questions remain unanswered. For whom is resilience built in sanctioned destinations, and at what social or ethical cost? Can adaptation under isolation ever be genuinely transformative, or does it simply normalize exclusion? How do power asymmetries, between states and communities, elites and workers, men and women, shape who bears the hidden burdens of resilience? What pathways might enable sanctioned destinations to re-enter global circuits, and how should scholars theorize processes of transition and reintegration? Finally, how do we, as tourism researchers, avoid becoming complicit in narratives that celebrate endurance while neglecting suffering?

This book does not claim to resolve these questions. Rather, it aims to open a space in which they can be asked, debated, and confronted. In doing so, the hope is that tourism scholarship can contribute not only to understanding resilience under sanctions but also to imagining more just, inclusive, and sustainable futures for destinations that continue to be caught in the crossfire of global politics.

References

Cortright, D., & Lopez, G. (2018). *Economic sanctions: Panacea or peacebuilding in a post-cold war world?* Routledge.

Demir, F., & Tabrizy, S. S. (2022). Gendered effects of sanctions on manufacturing employment: Evidence from Iran. *Review of Development Economics*, *26*(4), 2040–2069.

De Rogatis, P. (2022). Resilience in economic sanctions: The neglected relationship between resiliency and credibility in the EU. *Sentio*, *4*, 12–25.

Espinoza-Figueroa, F., Vanneste, D., Alvarado-Vanegas, B., Farfán-Pacheco, K., Rodríguez-Girón, S., & Saquicela, V. (2025). The misunderstanding between tourism resilience and survival: Stakeholder perceptions and policy effectiveness in Ecuador during the COVID-19 pandemic crisis. *Sustainability*, *17*(9), 4034.

Hall, C. M., & Seyfi, S. (2021). Tourism and sanctions. In *Research handbook on economic sanctions* (pp. 351–368). Edward Elgar Publishing.

Hove, M., Ndawana, E., & Nhemachena, M. A. (2020). How Cuba survived sanctions and the lessons for Zimbabwe. *Jadavpur Journal of International Relations*, *24*(2), 172–195.

Khodadadi, M. (2025). Tourism resilience in Iran: navigating sanctions, diplomacy and emerging opportunities. *Research in Hospitality Management*, *15*(2), 102–108.

Kozlovskaya, D. I., & Kozlovskaya, S. N. (2021). Patriotic tourism as a means education and upbringing of young people. *CITISE*, *4*(30).

Oghazi, P., Patel, P. C., & Hajighasemi, A. (2024). Gendered crisis approach: Exploring the gendered impact of Iranian sanctions on nascent entrepreneurship outcomes. *International Small Business Journal*, *42*(8), 1016–1046.

Rarick, C. A. (2006). Destroying a country in order to save it: The folly of economic sanctions against Myanmar. *Economic Affairs*, *26*(2), 60–63.

Rastegar, R., Seyfi, S., & Shahi, T. (2025). Tourism SMEs' resilience strategies amidst the COVID-19 crisis: The story of survival. *Tourism Recreation Research*, *50*(2), 428–434.

Seyfi, S., & Hall, C. M. (2019). International sanctions, tourism destinations and resistive economy. *Journal of Policy Research in Tourism, Leisure and Events*, *11*(1), 159–169.

Seyfi, S., Hall, C. M., Saarinen, J., & Vo-Thanh, T. (2023a). Do international sanctions help or inhibit justice and sustainability in tourism? *Journal of Sustainable Tourism*, *31*(12), 2716–2733.

Seyfi, S., Kimbu, A. N., Tavangar, M., Vo-Thanh, T., & Zaman, M. (2025). Surviving crisis: Building tourism entrepreneurial resilience as a woman in a sanctions-ravaged destination. *Tourism Management*, *106*, 105025.

Siyamiyan Gorji, A. S., Garcia, F. A., & Mercadé-Melé, P. (2023). Tourists' perceived destination image and behavioral intentions towards a sanctioned destination: Comparing visitors and non-visitors. *Tourism Management Perspectives*, *45*, 101062.

Zhang, J., Xie, C., & Huang, S. S. (2024). The effect of resilient leadership on employee resilience during a crisis in tourism & hospitality firms: A self-determination perspective. *International Journal of Hospitality Management*, *122*, 103886.

Index

For Product Safety Concerns and Information please contact our EU
representative GPSR@taylorandfrancis.com Taylor & Francis Verlag GmbH,
Kaufingerstraße 24, 80331 München, Germany

Printed and bound by CPI Group (UK) Ltd, Croydon, CR0 4YY
14/09/2026
02216243-0003